BUGS, BEETLES, SPIDERS & SNAKES

Ken Preston-Mafham, Nigel Marven & Rob Harvey

GREENWICH EDITIONS

A QUANTUM BOOK

This edition published by
Greenwich Editions
10 Blenheim Court
Brewery Road
London N7 9NT

ISBN 0-86288-364-4

QUMBBSS

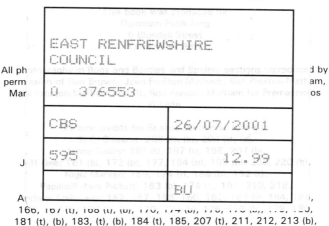

All photography in Bugs and Beetles and Spiders sections reproduced by
permission of Rod Brown, Jean Preston-Mafham, Ken Preston-Mafham,
Mar... Ken Preston-Mafham, Rod creator Mafham for Prema...os

...ure credits for Sn...

...ny Bailey: 164 (b), 197 (t), 198, 207 (b)
...(t), 172 (b), 177, 184 (b), 19...(t), 183 (t), 221 (b),
Nigel Marven...
...pilio/Robert Pickett: 163 (t), ...174, 187, 210, 218...
...y Amos: 152, 167, 16...17, ...162 (t), 184 (t) (b),
166, 167 (t), 168 (t), (b), 170, 174 (b), 176, 178 (b), 179, 180,
181 (t), (b), 183, (t), (b), 184 (t), 185, 207 (t), 211, 212, 213 (b),
214, 215 (b), 216, 217 (t), 220 (t), 221.
John Weigel: 156, 169, 171, 172 (t), 173, 175 (t), (b), 196 (t), (b),
199 (t), (b), 201 (t), (b), 202 (t), (b), 205 (b), 206, 219 (t), (b).

Material in this book previously appeared in *Bugs & Beetles Identifier* by Ken
Preston-Mafham, *Spiders Identifier* by Ken Preston-Mafham, *Snakes
Identifier* by Nigel Marven and Rob Harvey.

Typeset in Great Britain by
Central Southern Typesetters, Eastbourne
Manufactured in Singapore by Eray Scan Pte Ltd
Printed in China by Leefung-Asco Printers Limited

CONTENTS

BUGS

& BEETLES

**Over 100 species,
divided into two sections,
detailing the general characteristics
of each family.**

INTRODUCTION

Bugs and beetles can be found in every type of habitat, from gardens, parks, and woodlands to mountaintops and in lakes, ponds, and rivers. With their tough drought-resistant exteriors beetles are particularly well adapted to the hot arid conditions in deserts, where they are often the most abundant insects. Many species of both bugs and beetles are serious pests in gardens and on farmland. Others, such as ladybug beetles, are beneficial to mankind because of their depredations against such serious pests as aphids and scale insects.

BUG OR BEETLE?

The first question to be answered in using this book is how to tell a bug from a beetle. With the exception of a few shiny plump beetle-like bugs, the differences are quite obvious and, with just a little experience, can be seen at a glance, as explained below.

Bugs and beetles belong to two quite distantly related groups of insects which

Two instars of the Australian shield-backed bug *Tectocoris diophthalmus*, distinguished by differences in size and coloration. The two orange nymphs have just molted. They will soon harden and change to the brilliant metallic blue.

have very different methods of developing from egg to adult. Bugs belong to the order Hemiptera, and along with such familiar insects as grasshoppers, crickets, cockroaches, and earwigs undergo an "incomplete" metamorphosis which proceeds as follows. The juvenile stages are known as nymphs, which resemble miniature versions of the adults. These nymphs slowly develop through a series of molts and instars, during each of which tiny wing-pads on the outside of the body gradually grow until, after the final molt, the fully-winged adult (in winged species) emerges. In the majority of these families the nymphs utilize the same food as the adults.

Beetles belong to the order Coleoptera, and along with such insects as butterflies, bees, and flies, are considered to be more advanced insects on account of their "complete" method of metamorphosis. The creature which emerges from the egg is known as a larva. It is quite different than the adult and, unlike in most bugs, the adult and larval beetles do not necessarily share the

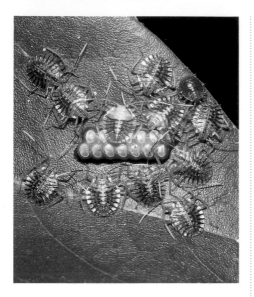

These *Edessa* sp. stink bug nymphs which have recently emerged from their eggs are just miniature versions of the adults. Their first meal will be from their own eggshells, from which they will pick up symbiotic protozoa which live in the gut. In many herbivorous bugs these play a vital role in digestion.

same source of food. Thus, whereas a longhorn beetle larva will have developed on a diet of wood inside a tree-trunk, the adult will feed solely on nectar and pollen taken from flowers. The larva goes through a series of molts until eventually, instead of changing finally into an adult, it enters a quiescent stage called a pupa. It is inside the pupa that the incredible process of metamorphosis takes place, during which the cells comprising the larval structure are broken apart and reassembled, like a complex jigsaw puzzle, to form the adult insect. After

a suitable interval the adult finally emerges from the pupa. In beetles whose larval diet is nutritionally poor, such as many wood-boring species, entire development from egg to adult can take many years to complete.

There is no simple rule for instantly distinguishing the immature stages of bugs and beetles. Bug nymphs always have legs, while many beetle larvae lack them, especially in darkling beetles (family Tenebrionidae) and weevils (family Curculionidae). If any doubt exists, then a quick inspection of the underside will reveal that a bug nymph will always have a needle-like feeding rostrum projecting downward from its head, or folded lengthwise beneath its body. By contrast, beetle larvae have biting mouthparts in which a pair of jaws normally figures conspicuously.

After crawling up a tree trunk at night, the nymph of this Australian cicada attaches its claws securely to the bark. A split then develops in the midline of the pronotum, through which the soft pallid adult gradually emerges.

Adults of the two groups can also usually be easily distinguished on this basis, but can also be told apart when seen from above and without any need for handling. In bugs, members of the suborder Homoptera, with their membranous wings held tent-like at an angle above the body, are quite unlike any beetles. In contrast, many members of the suborder Heteroptera could present problems, especially stink bugs (superfamily Pentatomoidea), which many people routinely confuse with beetles. The distinguishing feature is the way the wings meet above the back. In most stink bugs and other heteropterans the forewings, held flat against the body, meet in such a way that the membranous areas of the wing-tips overlap to form a

The larva of the tenebrionid beetle *Zophobas rugipes* is legless and very similar to that of the closely related meal-worm *Tenebrio molitor*, whose larvae are easily bred as food for various small animals.

The six-legged larva of the seven-spot ladybug *Coccinella 7-punctata* is quite unlike the black-spotted orange adult with its shiny domed body. Both adult and larva do however feed on aphids.

The pupa of the seven-spot ladybug is normally placed in full view on a leaf or stem. The molted skin of the larva is just visible on the right, at the point where the pupa is affixed to the leaf. By contrast, most beetle pupae are concealed from view.

triangle. A second triangle is normally formed by the scutellum. Bugs therefore normally have two triangles on their backs (see fig. 1).

In beetles the hardened wing-cases or elytra (actually modified front wings) meet together in a straight line down the middle of the back (see fig. 2). In most beetles the elytra are held upward out of the way so that the folded hindwings can be unfurled for flight. After landing, the hindwings are neatly folded away and the elytra closed. In the very beetle-like shield-backed bug (Scutelleridae) there is no line down the middle of the back where the wing-cases meet, thus indicating that it is not a beetle.

An adult leaf-footed bug (Coreidae) showing its stiletto-like sucking mouthparts (rostrum).

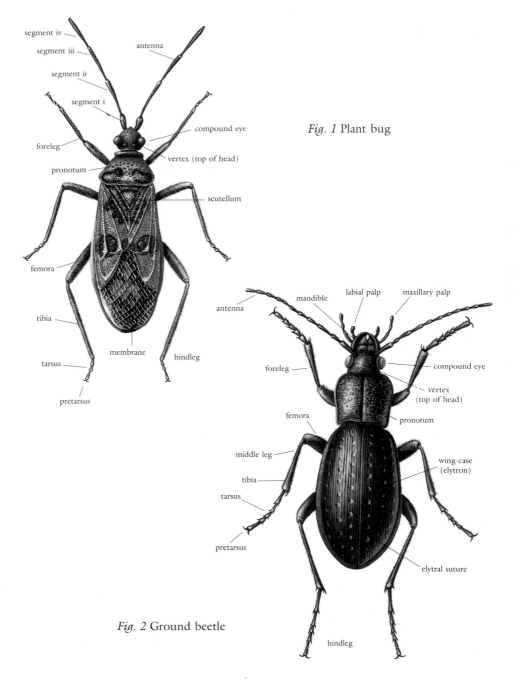

segment iv

segment iii

antenna

segment ii

segment i

Fig. 1 Plant bug

foreleg

compound eye

pronotum

vertex (top of head)

scutellum

femora

tibia

tarsus

membrane

hindleg

pretarsus

mandible

labial palp

maxillary palp

antenna

foreleg

compound eye

vertex (top of head)

femora

pronotum

middle leg

wing-case (elytron)

tibia

tarsus

pretarsus

elytral suture

Fig. 2 Ground beetle

hindleg

HOW TO USE THE IDENTIFIER

The first step is to decide whether or not you are dealing with a bug or a beetle, using figs. 1 and 2, in conjunction with the introduction. Then go through the photographs until you find the insect which closely matches the one you want to name. The information included within the "family" heading should then be consulted, along with the description of the insect you have chosen, to see whether or not your identification is likely to be correct.

The family tree below shows how the species are listed in order of classification, using their Latin and common names. We have selected three example species, the Anchor stink bug, the Peanut bug, and the Bee Beetle to show how the classification system works.

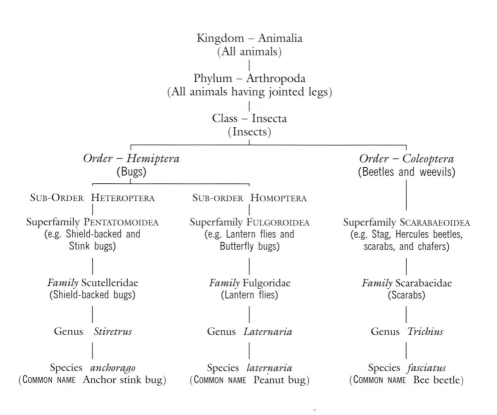

Kingdom – Animalia
(All animals)

Phylum – Arthropoda
(All animals having jointed legs)

Class – Insecta
(Insects)

Order – Hemiptera (Bugs)		*Order – Coleoptera* (Beetles and weevils)
SUB-ORDER HETEROPTERA	SUB-ORDER HOMOPTERA	
Superfamily PENTATOMOIDEA (e.g. Shield-backed and Stink bugs)	Superfamily FULGOROIDEA (e.g. Lantern flies and Butterfly bugs)	Superfamily SCARABAEOIDEA (e.g. Stag, Hercules beetles, scarabs, and chafers)
Family Scutelleridae (Shield-backed bugs)	*Family* Fulgoridae (Lantern flies)	*Family* Scarabaeidae (Scarabs)
Genus *Stiretrus*	Genus *Laternaria*	Genus *Trichius*
Species *anchorago* (COMMON NAME Anchor stink bug)	Species *laternaria* (COMMON NAME Peanut bug)	Species *fasciatus* (COMMON NAME Bee beetle)

Note that the genus and species are written in italics whenever they occur within the normal text.

The identifier section of this book is arranged in three groups, the sub-orders Heteroptera and Homoptera, within the Order Hemiptera (Bugs), and the Order Coleoptera (Beetles and Weevils). A brief description introduces each family, and each genus representing its family is identified with its species name, common name, together with information on its general characteristics and known distribution.

The symbols accompanying each entry convey essential information at a glance.

KEY TO SYMBOLS

LARVAL STAGES

 Larvae external, feeding visibly on leaves, stems, etc.

 Larvae are concealed, e.g. in timber, on roots, etc.

HABITAT

 Open areas generally, e.g. roadsides, grasslands, scrubland, etc.

 Forest and woodland, but often mainly in more open spots.

 Common in gardens and backyards.

 Mainly in deserts.

 In or beside water.

FOOD

 Herbivore, feeding on plants, e.g. leaves, stems, sap, wood, or fungi, etc.

 Predator, feeding on other small animals and insects.

 Scavenger, using dung, corpses, etc.

 Larvae are predators, but as adults the species is herbivorous.

BUGS
Order Hemiptera

Commonly known as bugs, this order of insects contains some 67,500 described species, split into two suborders, the Heteroptera and the Homoptera. The unifying feature of all bugs is their piercing and sucking mouthparts, housed in a beak-like rostrum.

SUBORDER HETEROPTERA

Known as the **true bugs** (to avoid confusion with inaccurate term "bugs" used to describe any insect-like creature), the members of this suborder possess a "hinged" rostrum which can be swung downward and forward from its stowed position along the underside of the body. The wings (if present) are held flat across the back when not in use, and the forewings exhibit a substantial hardened portion.

Superfamily ARADOIDEA

BARK BUGS, FLAT BUGS *Family Aradidae*

This is a medium-size family of some 1,800 species of very flattened bugs which usually live on or under the bark of trees, mainly those infected by fungi upon which the bark bugs feed. A few species make use of sap rather than fungi. Body size ranges from ⅛ to ½ inch, while the head is very characteristic in that the antennae stem from tubercle-like outgrowths.

DYSODIUS LUNATUS

COMMON NAME Bark bug

DESCRIPTION With a body length of about ½ inch this is one of the largest of the bark bugs. It often forms large aggregations on fallen trees in the American tropics. Note the very flattened body and the series of projections along the sides, that mimick a piece of flaking bark. Species from the USA and Europe are generally similar, though the latter are very much smaller.

DISTRIBUTION American tropical zones; common in rainforest in Costa Rica.

Superfamily CIMICOIDEA

DAMSEL BUGS *Family Nabidae*

Damsel bugs are quite slender dull-coloured insects which prowl slowly about among vegetation or on the ground, searching for small insects or spiders on which to prey. Just under 400 species are known worldwide, all being predators. During mating, the male penetrates the female's body-wall and the sperm makes its way through her body fluids toward the ovaries. Three species of *Nabis* in North America help control the boll worm *Heleothis zea*, a major pest.

DOLICHONABIS LIMBATUS

COMMON NAME Heath damsel bug

DESCRIPTION The body length ⅓–½-inch makes this a typical member of the family, although difficult to distinguish from other related species without reference to minute details of the body structure. It is found among the rank grass of damp meadows and in dense low vegetation bordering marshes. The adults are present from early July onward and are only rarely fully-winged. As with all damsel bugs, it will tackle any prey small enough to subdue. This one is feeding on a birch shieldbug nymph.

DISTRIBUTION Common throughout Europe, including the British Isles; many similar-looking species in North America.

Superfamily **PENTATOMOIDEA**

TORTOISE STINK BUGS *Family Plataspidae*

With their very rotund shiny bodies and flat undersides the members of this family may easily be mistaken for beetles. The scutellum is very large, enclosing most of the abdomen, but there is no line down the middle of the back, unlike in beetles. About 500 species have been described, chiefly from the warmer regions of the world. Most kinds feed on plants, but a few species make use of fungi.

LIBYASPIS COCCINELLOIDES

COMMON NAME Tortoise stink bug

DESCRIPTION In common with many members of the family, this species exhibits a marked difference between the adult and juvenile stages. The matte-finish nymphs are well-camouflaged against the bark of their host tree, while the ¾-inch long adults stand out and are very shiny and quite brightly colored. Note how the head of the adults is completely concealed beneath the huge pronotum.

DISTRIBUTION Tropical Africa and the island of Madagascar.

SHIELD-BACKED BUGS
Family Scutelleridae

The 400 or so members of this family are often mistaken for beetles. The scutellum is very large and extends across the entire abdomen, but without the beetle-like line down the middle of the back. Many species are brilliantly colored in metallic shades. All members of the family feed on plants and some have attained pest status.

STIRETRUS ANCHORAGO

COMMON NAME Anchor stink bug

DESCRIPTION The anchor-like marking is less obvious on the color-form pictured than on the equally common black and red form. This ⅓–½-inch long bug is a predator, feeding mainly on the soft-bodied larvae of various insects, but also on adult beetles as pictured. It mainly inhabits woodlands.

DISTRIBUTION USA and Central America.

TECTOCORIS DIOPHTHALMUS

COMMON NAME Harlequin bug

DESCRIPTION Males of this conspicuous bug attain a length of ¾ inch, while females may reach ⅞ inch. The scutellum is bright orange, patterned with scattered patches of iridescent blackish-blue or green. It feeds on various members of the hibiscus family, including cotton, on which it can become a pest. The female stands guard over her egg batch but does not stay with her offspring.

DISTRIBUTION Tropical and sub-tropical eastern Australia.

STINK BUGS
Family Pentatomidae

This is by far the largest family of stink bugs, with over 5,000 species worldwide. Most species have a rounded or broadly oval outline, although pointed projections on the sides of the pronotum are frequent. Sound production is quite common, being achieved by rubbing a row of pegs on the back legs against a ridged area on the underside of the abdomen. The great majority of pentatomids suck plant sap, but some common species are predatory. Odorous defensive secretions give rise to the common name.

ACANTHOSOMA HAEMORRHOIDALE

COMMON NAME Hawthorn shield bug

DESCRIPTION The berries of hawthorn trees (*Crataegus* sp.) form the principal food for both adults and nymphs, whose red and green coloration blends in well with the foodplant. The adults first appear in August and September, and then reappear in the following spring after hibernating through the winter. The body length of about ½ inch readily distinguishes it from the similar birch shield bug (*Elasmostethus interstinctus*) which is slightly smaller.

DISTRIBUTION Common and widespread in Europe, including the British Isles.

ELASMUCHA GRISEA

COMMON NAME Parent bug

DESCRIPTION The rather small female (about ⅓ inch) stands guard over her diamond-shaped egg-mass, usually laid on leaves, for around 2–3 weeks until the nymphs emerge. She then stands on sentry duty upon or beside her developing offspring for the next few weeks, until they are almost ready to molt into adults. The nymphs are much more colorful than the rather somber adult.

DISTRIBUTION Common on birch trees in Europe, including the British Isles.

GRAPHOSOMA ITALICUM

COMMON NAME Minstrel bug

DESCRIPTION With its longitudinal black and red stripes this ½-inch long bug can be mistaken for no other European species save *G. semipunctatum*, in which the pronotal stripes dissolve into dots. As will be obvious from the picture, the adults are quite different than the nymphs, which are a camouflaged shade of grayish-cream. The main foodplants are various umbellifers (carrot family).

DISTRIBUTION South and Central Europe, excluding the British Isles.

PALOMENA PRASINA

COMMON NAME Green shield bug

DESCRIPTION This is one of many similar-looking green shield bugs, but the only one present in the British Isles or common in northern Europe. The adults reach a length of ½ inch, feeding on a huge variety of plants, often in gardens. The slightly bigger green vegetable bug (*Nezara viridula*) is similar, but narrower in outline. It is common in warmer parts of the world, including southern Europe, the USA, and Australia, and has a striking pink, green, and black nymph.

DISTRIBUTION Found throughout many parts of Europe.

PERILLUS BIOCULATUS

COMMON NAME Eyed stink bug

DESCRIPTION As can be seen from this mating pair, the color of this striking bug is very variable. The markings can be red, orange, yellow, or white, but always set against a black background. This ⅓–½-inch long species is an important predator of the Colorado potato beetle.

DISTRIBUTION Widespread throughout most of the USA.

PEROMATUS sp.

COMMON NAME Banana bugs

DESCRIPTION This is one of two genera of large handsome stink bugs found in the American tropics, the other being *Edessa*. The adults reach a length of 1 inch or more, and are often boldly striped on the underside in black and yellow. The sides of the pronotum usually bear blunt projections. Many species feed on poisonous plants of the potato family *(Solanaceae)* and are occasionally introduced into some cooler regions of the world on shipments of tropical fruit, especially bananas.

DISTRIBUTION American tropical zones.

Superfamily LYGAEOIDEA

SEED BUGS *Family Lygaeidae*

This is a large worldwide family, with over 3,000 described species. The adult bugs are generally rather elongate-oval in outline, and the antennae project from low down on the head, below the eyes, When wings are present, five veins are clearly visible in the membranous hindmost portion of the front wings; many species have forms with vestigial wings. The rostrum consists of four segments. Some giants of the family reach a length of nearly ⅞ inch, but most are much smaller. The great majority of species feed on plants, but a few attack the eggs or immature stages of small insects or mites, while a few suck blood.

ONCOPELTUS FASCIATUS

COMMON NAME Large milkweed bug

DESCRIPTION Both the boldly marked ⅓-inch long black and orange adults and the bright red nymphs often congregate on milkweed plants, whose seeds comprise the main food. Particularly large overwintering groups can sometimes be seen on mild winter days.

DISTRIBUTION Eastern United States and Central America.

LYGAEUS KALMII

COMMON NAME Small milkweed bug

DESCRIPTION This handsome ⅓-inch long bug is red and black with a red mark on top of the head and two white spots on the wings. It is found on the flower heads and seed pods of various milkweeds. Two common European species are similar: *Lygaeus equestris* also has white spots on the wings, whereas *L. saxatilis* lacks them.

DISTRIBUTION Most of North America and down into Mexico.

NEACORYPHUS BICRUCIS

COMMON NAME Ragwort seed bug

DESCRIPTION At only ¼–⅓-inch this is rather smaller than the other common red and black seed bugs already mentioned. There is a conspicuous white cross on its back. It feeds solely on yellow-flowered ragworts such as *Senecio anonymus*. It can be difficult to find a lone bug, with mated pairs being the norm. This is because the males hang on to their mates in order to guard them against rivals until the next batch of eggs is laid, a daily event.

DISTRIBUTION Widely spread in eastern North America.

Superfamily **PYRRHOCOROIDEA**

COTTON STAINERS *Family Pyrrhocoridae*

This is a family of more than 300 species of bugs mainly found in the warmer parts of the world. Most species are warningly colored in red and black, especially in the nymphal stages, the adults often being rather drabber. The triangular head bears a long slender four-segmented rostrum, allied to four-segmented antennae.

DYSDERCUS sp.

COMMON NAME Cotton stainer bugs

DESCRIPTION This mating pair of just under ½-inch long cotton stainers in Mexico is representative of this large worldwide group of very similar-looking bugs. Even with expert knowledge it can be difficult to separate one species from another. The bright colors exhibited by both adults and nymphs warn that the bugs are unpleasant-tasting. The formation of large feeding aggregations of adults and nymphs is a common feature, with some species being pests of cultivated plants, including cotton.

DISTRIBUTION Mexico and southwestern regions of the USA.

PYRRHOCORIS APTERUS

COMMON NAME Firebug

DESCRIPTION The adults of this brightly colored ⅜-inch long bug are very seldom fully winged, usually occuring as short-winged. Its all-black head distinguishes it readily from various common red and black species of *Lygaeus*. After winter hibernation the adults often form huge conspicuous swarms on the ground. The firebug feeds on the seeds of a variety of plants, most especially members of the mallow and hibiscus families (Malvaceae).

DISTRIBUTION Much of Europe, especially in the south, extending to the Middle East. Very rare in the British Isles.

THICK-HEADED BUGS
Family Largidae

The 100 or so members of this family are very closely related to the Pyrrhocoridae, in which they are still placed by some experts. Largids mainly suck sap, but a few species are predacious. The family is heavily concentrated in the warmer parts of the world, where some species can reach 2 inches in length.

ARHAPE CICINDELOIDES

COMMON NAME Largid bug

DESCRIPTION This black and white ⅜-inch long species from the deserts of the southwestern United States is one of a number of largids which spend their lives among detritus on the ground. It rather resembles some of the tiger beetles with which it lives, hence the specific name, but it is more properly considered as an excellent mimic of some of the common black and white velvet ants (mutillid wasps) which pack a powerful sting.

DISTRIBUTION Deserts of southwestern USA and Mexico.

Superfamily **COREOIDEA**

LEAF-FOOTED BUGS *Family Coreidae*

The members of this family are sometimes known as squash bugs because one American species is a pest of cultivated squash. However, the rest of the 2,000-odd species feed on a wide variety of plants. The adults are invariably fully winged, and the membranous portion of the front wings bears a characteristic array of parallel veins. The four-segmented antennae are narrowed where they join the head. Some species attain a length of 1¾ inches, while many are brightly colored as nymphs but molt into drab brown or black adults. Stink glands are well developed.

NARNIA SNOWI

COMMON NAME Snow's leaf-footed bug

DESCRIPTION This rather small ⅓-inch long narrow-bodied species belongs to a genus typical of the deserts of the southwestern United States and Mexico. Here many of the commonest species feed on the fruits and stems of cacti. The species pictured feeds on a variety of plants, including yucca flowers (pictured) and juniper berries. Note the enlarged and flattened hind legs of this male, typical of the family. Most members of this genus are rather drably colored in grays and browns.

DISTRIBUTION Deserts of southwestern USA and Mexico.

ACANTHOCEPHALA TERMINALIS

COMMON NAME Common leaf-footed bug

DESCRIPTION The hind tibiae of this ⅜–⅞-inch long bug are expanded and leaf-like, unlike in the rather similar-looking true squash bug (*Anasa tristis*) which has cylindrical hind tibiae. The leaf-footed bugs feed on a wide variety of plants, and there are many broadly similar species.

DISTRIBUTION Over much of the USA and down into Mexico.

BROAD-HEADED BUGS
Family Alydidae

Once included with the Coreidae, the alydids are now widely treated as a separate family, being much more slender-bodied and elongate than the average coreid. The head is noticeably broad, much more so than in coreids, being more than half the width of the rear margin of the pronotum. The fourth antennal segment is always slightly curved and considerably longer than the previous segment. The nymphs of many species mimic ants. Just under 300 species are known from around the world.

HYALYMENUS sp.

COMMON NAME Ant bug

DESCRIPTION The ½-inch long adult pictured is typical of most alydids found in far distant parts of the globe. Note the enlarged and spiny back legs, typical of the genus, and slender build. The rostrum has been swung downward, ready to suck on the green fruit on which the bug's front legs are resting. The nymph is an excellent mimic of a large ant.

DISTRIBUTION Widely distributed in the warmer zones of the Americas. Similar-looking genera are found from Europe eastward to Australia and down into Africa.

SCENTLESS PLANT BUGS
Family Rhopalidae

All the known species of this 150-strong worldwide family feed on plants, sucking the seeds of a broad range of herbaceous kinds. In so doing, a few species make a nuisance of themselves and become pests. The body is often decorated with surface dimples, spines, and hairs. Some people still treat this family as part of the Coreidae.

LEPTOCORIS TRIVITTATUS

COMMON NAME Boxelder bug

DESCRIPTION The narrow brick-red markings down both sides of this ⅜–¾-inch long bug are characteristic. It is found in deciduous woodlands and in gardens, often on leaves of boxelder and other maples. However, it is also common on many other types of trees and shrubs, feeding on flowers and fruits. *L. rubrolineatus* from west of the Rockies is similar.

DISTRIBUTION Eastern North America.

CORIZUS HYOSCYAMI

COMMON NAME Red-backed bug

DESCRIPTION The superficial resemblance of this ⅓-inch long bug to some similar-size common black and red seed bugs can be confusing. However, this species is noticeably hairy and there are numerous veins in the forewing membrane (lygaeids never exceed five). The adults are usually found singly, sitting around on various plants. Adults which have survived the winter can be seen in June and July, with the new generation of adults appearing in late September.

DISTRIBUTION Widespread in Europe; rare in the British Isles.

Superfamily **TINGOIDEA**

LACE BUGS *Family Tingidae*

Tingids are mainly very small, flattened insects, with the largest being only ⅕ inch in length. Only under the microscope can the full beauty of the lace-like structure of the adults be appreciated. There are more than 1,800 species worldwide, all feeding on plants, with some becoming pests. A few species induce the formation of swellings, or galls, on their foodplants.

GARGAPHIA SOLANI

COMMON NAME Eggplant lace bug

DESCRIPTION This is one of several similar-looking species which are specialists on certain foodplants. The preferred foodplant of the eggplant lacebug is horsenettle (*Solanum carolinense*), but it also utilizes many other members of the potato family (Solanaceae). The females guard their own offspring, as well as those of other females who "dump" their eggs for others to foster. The adult guarding this mass of nymphs is at top-right of the picture.

DISTRIBUTION Widespread in North America.

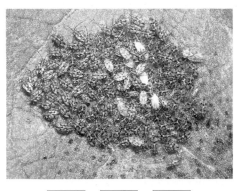

Superfamily **MIROIDEA**

PLANT BUGS *Family Miridae*

With 7,000–8,000 described species from around the world, this is by far the biggest of all the families of heteropteran bugs. Mirids are lightly built rather small bugs, the largest reaching ¾ inch in length. Antennae and rostrum are both four-segmented, while winged varieties exhibit two closed cells in the membranous region of the front wings. Although most species feed on plants, some are predatory and of value in controlling pests. Most bizarre are the select band which specializes in stealing prey from spiders' webs.

CALOCORIS STYSI

COMMON NAME Mirid bug

DESCRIPTION With a length of ¼–⅓ inch this is very much an average kind of mirid in size, but not in color, being one of the prettiest mirids in Europe. The yellow markings are diagnostic and cannot be confused with any other species. The eggs are laid in crevices in tree bark, and the resulting larvae eventually become adult in mid-June after feeding on a wide variety of plants, but especially on the developing catkins of stinging nettle *Urtica dioica*. Soft-bodied insects such as aphids also form part of the diet.

DISTRIBUTION Common throughout Europe, including the British Isles.

Superfamily **REDUVOIDEA**

AMBUSH BUGS *Family Phymatidae*

Often treated as a subfamily within the Reduviidae, the 100-odd known species of ambush bugs cannot be confused with any other heteropteran. Although usually quite small (about ⅕ inch), the body is bizarre in appearance, often being extravagantly decorated with flaps and other outgrowths from the sides. The front legs are highly modified for a raptorial lifestyle—they seize prey with their legs—forming a gin-trap much like the front legs of a praying mantis.

PHYMATA EROSA

COMMON NAME Common ambush bug

DESCRIPTION Like other ambush bugs, this species lies in wait on flowers to trap unwary insects in its spiny front legs. These are clearly visible in the above illustration of a mating pair, which were well camouflaged and almost invisible on the flowers they had chosen. The bug responds with lightning speed to the approach of a potential meal. The European species *P. monstrosa* and *P. crassipes* look very similar.

DISTRIBUTION Several similar-looking species over much of North America and Mexico.

ASSASSIN BUGS
Family Reduviidae

Over 5,000 species are known from this worldwide family of largely predacious bugs. A few species suck blood and in so doing may affect man and domestic animals. Body length varies from ⅓–1⅝ inches, while shape varies from short and relatively stout to long and almost thread-like. The head is very narrow, while the rostrum has three segments.

APIOMERUS FLAVIVENTRIS

COMMON NAME Bee assassin bug

DESCRIPTION Numerous similar-looking species of this genus wait in ambush on flowers in North America. Sometimes the bug is a good color-match for the flower, but this does not seem to be necessary in order to catch prey, as seen here with a bug having caught a bee on a contrasting yellow flower. Reduviids are bold predators, ready to tackle large fierce prey. The rostrum injects a powerful fast-acting toxin which invites care in handling, as the effect on a human finger can be more painful than a bee or wasp sting.

DISTRIBUTION Western North America and Mexico, with many similar-looking species.

PSELLIOPUS ZEBRA

COMMON NAME Zebra assassin bug

DESCRIPTION As in most species of *Pselliopus*, the legs, antennae, and abdomen of this ½-inch long bug are banded in black and creamy white. These bugs are found sitting or walking on foliage, rather than on flowers. They prey on a wide variety of small insects and spiders. *P. cinctus* from the eastern USA differs mainly in having a plain brownish-red top to the pronotum.

DISTRIBUTION Mexico.

Superfamily GERROIDEA

WATER STRIDERS or POND SKATERS *Family Gerridae*

Members of this family are conspicuous elements of the surface life of ponds, lakes, and rivers, and even the open ocean, certain gerrids being the only insects to effect successful colonization of this environment. The 500-odd known species are found throughout the world, the majority being wingless, although winged forms do arise when overcrowding forces a move to new waters. Gerrids move easily over the surface film on their middle and hind legs, which are long and slender and have unwettable feet. The body is also furnished with a dense felt of unwettable hairs.

GERRIS LACUSTRIS

COMMON NAME Common pond skater

DESCRIPTION With a length of ⅛–⅜ inch, this is an average member of the family, with the drab appearance typical of the group. In common with all gerrids, it can detect stranded prey struggling on the water surface, using a combination of acute eyesight and ripple-sensitive hairs on the body and legs. Thus, several gerrids will quickly converge on a floundering damselfly. Cannibalism also occurs, especially toward young nymphs. There are many similar-looking species in North America (e.g. *G. remigis*). Here the pond skater is skating on duckweed.

DISTRIBUTION Widespread in Europe and most of Asia.

Superfamily **GELASTOCOROIDEA**

TOAD BUGS *Family Gelastocoridae*

Like their namesakes, toad bugs have a squat warty appearance and proceed via a series of hops. The very flattened face is surmounted by two large bulging eyes, while the short four-segmented antennae are normally held out of sight beneath the eyes. All the 85 or so species are predatory, and no species exceeds ½ inch in length.

GELASTOCORIS PERUENSIS

COMMON NAME Toad bug

DESCRIPTION Like most toad bugs, this brown, warty ⅓-inch long species runs around on the borders of ponds and streams and in rainforests, where it blends in well with its surroundings, resembling a small pebble. The front legs are modified for snapping up prey, comprising smaller insects. Numerous similar-looking species are found throughout the Americas e.g. the western toad bug (*G. variegatus*) from the western USA and Mexico, and the big-eyed toad bug (*G. oculatus*) from most of the USA and Mexico.

DISTRIBUTION Streamsides and ponds in tropical America.

Superfamily NOTONECTOIDEA

BACKSWIMMERS *Family Notonectidae*

Backswimmers have a rather boat-like body and strong oar-like hind legs which are fringed with hairs. These propel the insect, on its back, in a series of jerks through the water. Although aquatic, backswimmers make regular visits to the water surface, gathering air and storing it beneath the wings and on special hairs on the underside of the abdomen. More than 300 species are known worldwide.

NOTONECTA GLAUCA

COMMON NAME Common backswimmer

DESCRIPTION The bug pictured has come to the surface and is lying briefly on its back in order to gather an air supply. When it submerges, it will have to row powerfully downward, as it is now lighter than water. Like all water boatmen, it is a ferocious predator, using its powerful beak-like rostrum to stab small prey such as damselflies, small fish, and tadpoles. If carelessly handled, the result can be a very painful stab in the finger. Its length is usually about ½ inch.

DISTRIBUTION Many similar-looking species around the world. Commonest North American species is *N. undulata*.

SUBORDER HOMOPTERA

This suborder differs from the Heteroptera in the following ways. The rostrum or beak cannot be directed forward to feed and the wings are held tent-like above the body when at rest. More than 42,500 species are known, many of which produce honeydew as a waste product.

Superfamily FULGOROIDEA

LANTERN FLIES *Family Fulgoridae*

This family of about 750 species contains some of the world's most bizarre looking insects, especially in the shape of the head, which is often modified in all kinds of strange ways. There are no really small species, the smallest being ⅜ inch in length, while the largest can reach 4 inches or more. The family reaches its maximum diversity in the tropics, especially in rainforest areas.

LATERNARIA LATERNARIA

COMMON NAME Peanut bug

DESCRIPTION This spectacular bug is one of the giants of the family, reaching a length of 3⅕ inches with a wing-span of 4 inches. The common name is derived from the huge bulbous projection on the head which resembles an unshelled peanut. This head was once erroneously thought to be luminous, hence the common name of lantern flies given to the whole family. The peanut bug sucks the sap from the trunks of large trees, and when disturbed opens its wings to reveal intimidating eyespots.

DISTRIBUTION Over much of the American tropics and subtropics.

BUTTERFLY BUGS
Family Flatidae

The common name given to these bugs is derived from some of the larger and very colorful tropical species, which resemble butterflies when in flight. More than 1,000 species are found around the world, mainly in tropical and subtropical zones. The broad front wings are never transparent and have blunt ends, being held at a steep angle above the body when at rest.

PHROMNIA ROSEA

COMMON NAME Flower-spike bug

DESCRIPTION Both the adults, which are about ⅞ inch long, and the strange white nymphs form large feeding groups on the branches of small trees. The adult bug can be either vibrant pink or green, and when assembled the mass of bugs bears a remarkable resemblance to an attractive spike of brightly colored flowers. The nymphs have long waxy white "tails," a distinctive and common feature in the family.

DISTRIBUTION Tropical Africa and the island of Madagascar.

FALSE LANTERN FLIES
Family Dictyopharidae

Although widely distributed around the world, this family only contains some 500–600 species of mostly green or brown bugs. The head is prolonged forward in a long snout-like projection called the cephalic horn. This rather resembles the head of some smaller lantern flies, hence the common name. Most species feed on grasses.

EPIPTERA EUROPAEA

COMMON NAME European lantern fly

DESCRIPTION This bright green ⅜-inch long bug is found on a broad range of herbaceous plants, particularly umbellifers (carrot and parsley family). The adults can be found from the end of June through October, often only singly. The complex network of veins toward the tips of the transparent wings is diagnostic. In central Asia this species can be a problem in melon plantations. The common name is misleading, as this bug is not a member of the Fulgoridae.

DISTRIBUTION Europe eastwards to central Asia; absent from the British Isles.

Superfamily CERCOPOIDEA

FROGHOPPERS *Family Cercopidae*

With over 2,500 species mainly around warmer zones of the world, these bugs are usually brown, able to jump, and have a frog-like appearance. The front wings are longer than the body and the tibiae are round in cross-section. They feed on plant sap.

CERCOPIS VULNERATA

COMMON NAME Black and red froghopper

DESCRIPTION One of the few colorful members of the family. The ⅜-inch long adults are conspicuously present from late April to June, sitting around on a wide variety of herbaceous vegetation. The nymphs form feeding aggregations on the roots of grasses, surrounded by a ball of sticky froth. Several similar species are present in Europe.

DISTRIBUTION Widespread in Europe, including the British Isles.

PHILAENUS SPUMARIUS

COMMON NAME Meadow spittlebug

DESCRIPTION In this species the nymph makes its presence far more obvious than the adult. Sap excreted from the feeding nymph's rear end forms a froth which covers its soft delicate body. The ¼-inch long adult occurs in numerous color-forms in shades of brown, black, and gray.

DISTRIBUTION Widespread in Europe, including the British Isles, and Asia; eastern and western fringes of North America.

Superfamily **CICADOIDEA**

CICADAS *Family Cicadidae*

In most warmer parts of the world cicadas are the most obvious of all homopterans, due to the considerable volume of noise made by the calling males. There are some 2,250 described species, ranging from a little under ⅜ inch to giants over 4 inches long. The wide blunt head is distinctive, as are the large usually transparent wings, which make cicadas strong fliers. Most species are brown, gray, green, or blackish, but some tropical kinds are beautifully colored.

TIBICEN CANICULARIS

COMMON NAME Dogday harvestfly

DESCRIPTION This 1–1⅛-inch long species inhabits coniferous and mixed woods, appearing on the wing from late summer to early fall. The nymphs live below ground, sucking the roots of pine trees, taking three years to reach the adult state. The adults emerge synchronously so that locally trees and bushes will be almost covered in them.

Males "sing" by rapidly clicking in and out a tymbal membrane, amplified by air sacs within the body, producing a sound like a circular saw cutting through timber.

DISTRIBUTION Northeastern USA and adjacent region of Canada.

Superfamily CICADELLOIDEA

LEAFHOPPERS *Family Cicadellidae*

This is a huge worldwide family with over 20,000 species, making it far and away the largest family within the Hemiptera. They are slender-bodied insects with a noticeable taper toward the rear end. Most species measure less than ¾ inch from head to wingtip, and even the tropical "giants" do not go much over ⅞ inch. The hind tibiae are distinctive with their angular and rather flattened cross-section, adorned with rows of spines. Leafhoppers suck plant sap, with few if any plant species escaping their attentions.

CICADELLA VIRIDIS

COMMON NAME Green leafhopper

DESCRIPTION Adults of this species are usually shades of brown, green, or yellow, although the males are often a distinct bluish shade. Females reach ⅓ inch in length, the males only just over ¼ inch. They occur from July to September on vegetation in marshy areas, where the eggs are laid in the leaves and stems of rushes. The vampire leafhoppers (*Draeculacephala* sp.) found all over North America have larger and more pointed heads, but are otherwise similar. The rather bigger redbanded leafhopper (*Graphocephala coccinea*), from the eastern USA, is similar but has longitudinal red striping.

DISTRIBUTION Widespread in Europe, including the British Isles.

Superfamily **MEMBRACOIDEA**

TREEHOPPERS *Family Membracidae*

This is a large worldwide family containing nearly 2,500 species. Small size is normal, with about ¾ inch being the maximum body length, but numerous species form conspicuously large aggregations, both as adults and nymphs. The distinguishing feature to look out for is the considerable extension of the pronotum, which in many tropical species assumes complex and bizarre shapes.

CYPHONIA CLAVATA

COMMON NAME Ant treehopper

DESCRIPTION This ⅓-inch long treehopper is one of a number of species from the American tropics with pronotal adornments, thought possibly to be mimicking an open-jawed ant. As few predators feed on ants, this would be a reliable form of protection. The transparent wings enable the green body to blend into the leaf, strengthening the ant-like appearance. This species occurs singly as adults on low vegetation on forest edges.

DISTRIBUTION American tropics.

UMBONIA CRASSICORNIS

COMMON NAME Horned treehopper

DESCRIPTION The smaller bug (⅜ inch) with the elongate vertical blackish extension to the pronotum is this species. Facing it is an adult of *Umbonia spinosa*, with a shorter pronotal "thorn." Neither species actually mimics thorns, and seldom if ever lives on thorny trees. Both species are warningly colored, with the thorn-like projection making them difficult to swallow, giving their bad-tasting properties time to invoke rejection before any harm is done.

DISTRIBUTION Widely spread in the American tropics and subtropics, with *U. crassicornis* just getting into Florida.

Superfamily **APHIDOIDEA**

APHIDS *Family Aphididae*

Other common names for this large family of more than 4,000 species of tiny bugs are greenfly, blackfly, and plantlice. The giants among aphids only reach a body length of ¼ inch, while some species are only about ¹⁄₂₀-inch long. The most distinctive structural feature is a pair of tube-like cornicles or siphunculi which arise from the the fifth or sixth abdominal segments. These cornicles produce defensive secretions which entrap enemies, such as parasitic wasps. Many species have generations which alternate on different foodplants.

MACROSIPHUM ALBIFRONS

COMMON NAME Lupin aphid

DESCRIPTION Unlike some of its close relatives, the lupin aphid does not alternate its hostplant, but sticks solely to various lupins. This gray-green mealy-looking, farinacious species can be a major pest of cultivated lupins, especially in Europe where it is of recent introduction. The plump female at top right is just giving virgin birth (known as parthenogenesis) to a live nymph, the normal method of aphid reproduction in summer. In the fall, aphids mate and lay normal eggs.

DISTRIBUTION Virtually worldwide where lupins are cultivated or native.

Superfamily **COCCOIDEA**

MEALY BUGS *Family Pseudococcidae*

There are more than 1,000 species in this family distributed worldwide. Many of them are serious pests of cultivated plants and are difficult to control with chemicals. This is partly because many species have a waxy coating which often gives good protection against chemical attack. Body length is generally less than ⅛ inch. Reproduction in many species is parthenogenetic, and males are not known.

PSEUDOCOCCUS AFFINIS

COMMON NAME Glasshouse mealybug

DESCRIPTION This worldwide pest of cultivated plants in glasshouses has the waxy-white appearance typical of the group and from which the common name is derived. Like in all mealybugs, the female is wingless, but is able to walk considerable

distances to colonize new plants. The female's egg-sac is abundantly covered in white waxy filaments and covers her body.

DISTRIBUTION Worldwide in glasshouses.

BEETLES
Order Coleoptera

The Coleoptera is the largest order of insects with well over 300,000 known species. The front wings or elytra are tough and horny and in most species cover the entire abdomen. The membranous hindwings (when present) are neatly folded beneath the elytra when not in use. The mouthparts are normally of the biting type, rather than adapted for sucking as in bugs.

Superfamily **CARABOIDEA**

TIGER BEETLES and GROUND BEETLES *Family Carabidae*

This is a huge worldwide family of mainly shiny predatory beetles which are usually fast and agile runners on the ground. All kinds of prey are taken, but some species are specialists on certain types. The jaws are powerful and efficient, while the antennae are slender and usually have eleven segments. Some species can reach a length of 3 inches.

CARABUS VIOLACEUS

COMMON NAME Violet ground beetle

DESCRIPTION Only in certain lights is the violet shimmer visible on the elytra of this common beetle. It varies between ⅞ and 1⅜ inch and the very smooth shiny elytra help separate it from related species which have grooved or pock-marked elytra (e.g. the European ground beetle (*C. nemoralis*) which is also common in the USA and Canada). The violet ground beetle is found in many habitats, including gardens, but is seldom seen in daytime, emerging at night to hunt slugs and other prey.

DISTRIBUTION Common over much of Europe, including the British Isles, eastwards to Siberia and Japan.

CICINDELA SEDECIMPUNCTATA

COMMON NAME 16-spot tiger beetle

DESCRIPTION This ½-inch long species is one of a group of tiger beetles from the USA in which individual species are differentiated by their varying number of spots. Tiger beetles are active on warm sunny days, running rapidly across bare open ground on their long legs. They tear prey apart with their powerful curved mandibles.

DISTRIBUTION Southwestern USA.

CICINDELA SEXGUTTATA

COMMON NAME 6-spotted green tiger beetle

DESCRIPTION The body, antennae, and legs of this ½-inch long species are a scintillating shade of bluish green. The number of white spots is variable, there often being only 3–4, or sometimes none at all. It is one of only a few tiger beetles which live in forests, where it can be found running on paths or in open areas.

DISTRIBUTION Eastern USA and adjacent areas of Nova Scotia.

CICINDELA LANGI

COMMON NAME Lang's tiger beetle

DESCRIPTION This ¾-inch long beetle is one of 90 or so members of the genus found in the USA. Most of the species, including this one, are found in the west, being characteristic insects of flat sandy areas in deserts. Tiger beetle larvae are strange creatures with huge heads which are used to block a vertical burrow. When an insect strays close, the waiting larva grabs it.

DISTRIBUTION Southwestern USA.

Superfamily STAPHYLINOIDEA

BURYING BEETLES or CARRION BEETLES *Family Silphidae*

Also called undertaker beetles, this family of over 2,000 species is widely distributed, especially across the temperate zones of the northern hemisphere. The antennae are characteristically club-shaped or have a thickened button-like tip; the hind part of the abdomen is exposed. Not all species live on carrion; some are predators, while others live on vegetable debris or fungi.

NICROPHORUS VESPILLO

COMMON NAME Burying beetle

DESCRIPTION This is one of several similar-looking burying beetles having orange bands on the black elytra. The terminal antennal segments are reddish-brown (black in the very similar *N. vespilloides*); body length is ½–⅞ inch. All members of the genus exhibit advanced parental care, burying small corpses for use as a food store in the rearing of their larvae. Male and female often cooperate in this domestic task, and communicate with their offspring through chirps.

DISTRIBUTION Found throughout the whole of temperate Europe and Asia, and across most of North America.

ROVE BEETLES
Family Staphylinidae

This comprises a huge worldwide family of beetles with over 20,000 known species. In general they are easily recognized by their very short elytra, usually much shorter than the body. However, a few species have fairly long elytra, while members of some other unrelated families also have the elytra much abbreviated. Most rove beetles are predators.

ONTHOLESTES TESSELLATUS

COMMON NAME Tessellated rove beetle

DESCRIPTION The body of this ½–¾-inch long species is liberally sprinkled with short golden hairs, which make the beetle scintillate as it moves in its characteristic jerky fashion. It can be found, often in large aggregations, on fresh dung and carrion, in which the eggs are laid. The gold-and-brown rove beetle (*O. cingulatus*) from across the whole of North America is very similar.

DISTRIBUTION Common throughout Europe, including the British Isles, and eastward into Asia.

Superfamily **CANTHAROIDEA**

NET-WINGED BEETLES *Family Lycidae*

Lycids are noticeably flattened beetles, usually with very narrow beak-like heads and robust many-segmented antennae. The rather soft pliable wing-cases are decorated with a conspicuous lattice-work of veins. In some species the mouthparts are modified to form a nectar-sucking proboscis. The larvae mostly live in rotting wood where they prey on other insects. Distribution is worldwide, with a bias towards the warmer areas.

LYCUS ARIZONICUS

COMMON NAME Arizona net-winged beetle

DESCRIPTION The ½-inch long species pictured is very similar to members of the genus found in many parts of the world. All lycids deploy distasteful defensive fluids. Beetles of many other families, plus a few moths, mimic lycids, making it important to make a careful inspection of any beetle thought to belong to this family. The end-band netwing (*Calopteron terminale*) is very similar. It is found throughout North America, but lacks the long "snout" found in species of *Lycus*.

DISTRIBUTION Southwestern regions of the USA and Mexico.

GLOWWORMS or FIREFLIES
Family Lampyridae

Some 2,000 species are found throughout the world, especially in the tropics. After dark winged males fly in search of the stationary wingless often larva-like females, flashing a light which acts as a sexual message to the females. Many females can respond with their own luminous reply. The light is produced chemically.

LAMPYRIS NOCTILUCA

COMMON NAME Common glowworm

DESCRIPTION The adult male, a ½-inch long flattened and rather narrow-bodied brown insect is seldom seen. He is not luminous, but flies around at night looking for the greenish light emitted from the rear underside of the brown larva-like females. Most often seen are the orange-spotted larvae (pictured), which may be found under stones or logs, or walking around among grass. They feed mainly on snails.

DISTRIBUTION common in central and southern Europe, including England.

SOLDIER BEETLES
Family Cantharidae

Cantharids are rather small narrow-bodied beetles with unusually soft elytra. They derive their name from the bright colors of some species, which can resemble military uniforms. The adults are generally predatory, searching for prey on flowers, but they also eat nectar and pollen. Many species are warningly colored. The larvae hunt mainly on the ground.

CANTHARIS RUSTICA

COMMON NAME Black and red soldier beetle

DESCRIPTION This ⅓–½-inch long species is one of more than 15 similar-looking species found in Europe. It is common on foliage in early summer, and is often found mating, a process which can be very protracted. The downy leather-wing *(Podabrus tomentosus)* looks very similar; it is common over the whole of North America.

DISTRIBUTION Widespread in Europe.

SOFT-WINGED FLOWER BEETLES
Family Melyridae

These generally rather small beetles resemble small soldier beetles with similarly soft elytra but much more tapered rear ends. However, the relationship between the two families is not close. The adults are generally covered with long hairs or scales, and are mainly found on flowers. The larvae are predatory, mostly living in rotten wood.

MALACHIUS BIPUSTULATUS

COMMON NAME 2-spotted flower beetle

DESCRIPTION This ¼-inch long species is easily recognized by its combination of small size and green elytra, marked with red only at their tips. It is usually seen during May and June on flowering grasses, where it spends a considerable time feeding on the pollen. The males have yellowish outgrowths called excitators on the antennae, on which the female feeds during courtship.

DISTRIBUTION Common over much of Europe, including the British Isles.

Superfamily CUCUJOIDEA

FLAT BARK BEETLES *Family Cucujidae*

Members of this family can be easily recognized by their very flattened, rather elongate parallel-sided body, although a few species are not noticeably flattened. They are usually found walking around on logs in shady forests, or can be located by searching under loose bark. Most species feed on mites and small insects, while a few can be pests of stored products such as grain.

HECTARTHRUM GIGAS

COMMON NAME Flat bark beetle

DESCRIPTION This ½–¾-inch long species can be found on damp fungus-ridden logs in many African forests. The jaws are noticeable from above, while the antennae, which resemble a string of beads, are very characteristic. The red flat bark beetle species *(Cucujus clavipes)* is distinguished by its shorter, broader pronotum and all-red body. It is found throughout North America.

DISTRIBUTION East and Central Africa.

FUNGUS BEETLES
Family Erotylidae

With their generally rather domed shiny bodies, fungus beetles could be easily mistaken for certain kinds of leaf beetles. The clubbed antennae arise in front of or between the eyes, while the feet terminate in broad, hairy pads. Erotylids feed on fungi both as larvae and adults, and some tropical species pupate in huge aggregations on fungus-infected logs in rainforests. At least one species exhibits advanced parental care.

MEGALODACNE HEROS

COMMON NAME Pleasing fungus beetle

DESCRIPTION The bright colors of this ½-inch long beetle are certainly pleasing, hence the common name. It is usually found in deciduous forests, where the females may be located laying their eggs in a fungus-infected tree.

DISTRIBUTION Forests of eastern USA.

HANDSOME FUNGUS BEETLES
Family Endomychidae

The endomychids are exclusively small beetles which rather resemble ladybugs. However, endomychids are much flatter and have much longer antennae. The form of the pronotum is also characteristic, with two lengthwise grooves at its base. Both larvae and adults are found on fungi, decaying wood, and rotting fruit.

ENDOMYCHUS COCCINEUS

COMMON NAME Handsome fungus beetle or false ladybug

DESCRIPTION This ⅕–¼-inch long beetle greatly resembles one of the smaller species of ladybugs. The larvae are also brightly colored and feed openly on fungi and fungus-infested wood, although also occurring under the bark. The adults are seen from April to June. *E. biguttatus* from North America is very similar, but has a black pronotum.

DISTRIBUTION Widespread in Europe, including the British Isles.

LADYBUG BEETLES or LADYBIRDS
Family Coccinellidae

This is perhaps the best known and well-liked family of beetles, called ladybirds in Europe. It has some 3,400 species around the world. Ladybugs are easily recognized by their domed bodies with flattened undersides, bright colors, spots, and small head which is retracted into the pronotum.

COCCINELLA NOVEMNOTATA

COMMON NAME Nine-spotted ladybug

DESCRIPTION This ¼-inch long species is easily recognized by the nine black spots on an orange background, four spots on each elytron, and one on the scutellum. The head and thorax are black and have yellowish or whitish marks on the margins. It is found in meadows, gardens, parks, fields, and marshy places.

DISTRIBUTION Throughout North America except for the southwest.

COCCINELLA SEPTEMPUNCTATA

COMMON NAME Seven-spot ladybug

DESCRIPTION At ¼–⅓-inch long, this is one of the larger European ladybugs, and perhaps the commonest and most familiar. It is easy to count the seven black spots on the individual on the left of the illustration. The smaller ladybug on the right is a ten-spot ladybug (*C. 10-punctata*). Both species are basking in the light after emerging from winter hibernation.

DISTRIBUTION Common throughout Europe, including the British Isles; introduced into the USA and now common throughout most of the northeastern states.

ADALIA BIPUNCTATA

COMMON NAME Two-spot ladybug

DESCRIPTION Many ladybugs are notorious for the variability of their pattern, but none more so than the ⅕-inch long two-spot ladybug. The degree of variation is evident from the accompanying illustration of a mating pair. Every possible combination of red and black occurs. As in most ladybugs, both larva and adult feed on aphids.

DISTRIBUTION Europe, the British Isles, and throughout the whole of North America.

HIPPODAMIA CONVERGENS

COMMON NAME Convergent ladybug

DESCRIPTION This ¼–⅓-inch long species, like many ladybugs, shows great variation, particularly in the number of black spots. There are usually 13 spots, six on each elytron and one on the pronotum. However, their number can be reduced to two or three or even zero. The two converging white stripes on the pronotum are diagnostic. Huge overwintering aggregations in the mountain forests of the west, as seen here, are typical.

DISTRIBUTION Throughout North America.

CARDINAL BEETLES or FIRE COLORED BEETLES
Family Pyrochroidae

This is a small family of mainly reddish-brown rather flattened beetles with a superficial resemblance to some net-winged beetles (Lycidae). The antennae are often pectinate or shaped like a comb. The larvae develop under bark where they attack other insects.

PYROCHROA SERRATICORNIS

COMMON NAME Red-headed cardinal beetle

DESCRIPTION This is one of two similar-looking ½–¾-inch long species which are common in Europe. The other species *P. coccinea* is easily distinguished by its black head and brighter red coloration. Both species are commonly seen on flowers such as buttercups in May and June on the edges of woodlands or along shady lanesides.

DISTRIBUTION Found over much of Europe, including the British Isles.

FALSE BLISTER BEETLES
Family Oedemeridae

A small worldwide family of beetles which, with their slim outline, could easily be confused with soldier beetles (Cantharidae). In oedemerids the antennae are long and thin, while the rather soft elytra are strongly ribbed and often taper to a point, leaving a gap between them toward the rear. The larvae develop inside rotten wood and within the dry stems of herbaceous plants.

OEDEMERA NOBILIS

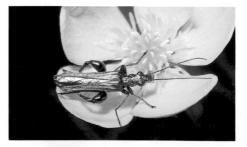

COMMON NAME Thick-legged flower beetle

DESCRIPTION This is one of several species in both *Oedemera* and *Oncomera* in which the hind femora of the male are grossly swollen. At ⅓–⅜ inch, the males are slightly larger than the normal-legged females. The adults are often abundant in June and July on flowers, and fly readily.

DISTRIBUTION Widespread but scattered over much of Europe, including the British Isles.

Superfamily **CLEROIDEA**

CHEQUERED BEETLES *Family Cleridae*

Some 3,000 species of these brightly colored rather hairy beetles occur around the world, although the great majority of them are in tropical areas. The larvae are normally predatory, often found feeding in the nests of bees and wasps, earning them the name of bee wolves.

TRICHODES ORNATUS

COMMON NAME Ornate chequered beetle

DESCRIPTION The elytra of this ⅕–¾-inch long beetle can be marked with red and black, as seen here, or equally commonly with yellow and black. The adults are usually found on flowers, where they feed both on the pollen and on other small insects. The larvae are found within the nests of solitary bees and wasps, where they prey upon the brood.

DISTRIBUTION Throughout the whole of North America.

Superfamily ELATEROIDEA

CLICK BEETLES *Family Elateridae*

The 8,000 or so species of click beetles populate a large part of the earth's land surface. Body length varies from around ⅕ inch up to 3½ inches in some tropical giants, some of which are also luminous. The body is characteristically long and rather narrow, tapering somewhat at the rear end. Some of the large tropical species are so hard that it is impossible to hammer a pin through without bending it.

CHALCOLEPIDIUS ZONATUS

COMMON NAME Harlequin click beetle

DESCRIPTION This large 2-inch long click beetle is usually found on fallen trees in forests. The margins of the pronotum bear two broad white bands, while the elytra are marked with a series of longitudinal black and white stripes. The black and white click beetle (*C. webbii*) from southwestern North America is slightly smaller (about 1½ inches) but otherwise very similar, save for its shiny black elytra on which the white is restricted to the margins.

DISTRIBUTION Widespread in South America.

SEMIOTUS AFFINIS

COMMON NAME Elegant click beetle

DESCRIPTION This is one of many large and spectacularly colored elaterids found in the tropics. The 1⅕-inch long female can often be found walking around on fallen trees on which she lays her eggs. Click beetles derive their name from the spring device on their undersides, which can propel them a few inches into the air when they are lying on their backs.

DISTRIBUTION Widespread in rainforests of the American tropics.

Superfamily **BUPRESTOIDEA**

JEWEL or **SPLENDOUR BEETLES** *Family Buprestidae*

Although there are some 15,000 species of buprestids around the world, the vast majority of them, and all the really large species, are restricted to the tropics. They are active in sunshine and have a variety of bright metallic colors. They range in size from under ⅛ inch up to 3 inches. The antennae are generally rather short. The larvae normally develop inside timber.

STIGMODERA RUFOLIMBATA

COMMON NAME 6-spotted jewel beetle

DESCRIPTION This rather small species (½ inch) can be quite common on *Melaleuca* and *Leptospermum* flowers in late spring. Note the bullet-shaped body, blunt head, antennae folded away, and scarcely visible legs, all typical features of the family. Most members of this genus are warningly colored, advertising the presence of bitter defensive chemicals called buprestins.

DISTRIBUTION Western and southern Australia, in open bushland and sandplains.

JULODIS HIRSUTA

COMMON NAME Hairy jewel beetle

DESCRIPTION This species, just over 1 inch long, is one of several abundantly hairy members of the genus occurring in southern Africa. These therefore lack the normal metallic sheen typical of the family. Various metallic species of the genus are found from the Cape region of South Africa north through Africa as far as the Middle East.

DISTRIBUTION Southern Africa, in dry areas.

Superfamily **MELOOIDEA**

OIL BEETLES or BLISTER BEETLES *Family Meloidae*

Most of the beetles in this family are brightly patterned in warning colors. These warn would-be enemies that the beetles should not be touched, as they can release an evil-smelling poisonous liquid from their joints when molested. The larvae are parasitic on other insects. More than 2,300 species occur worldwide, mainly in warmer areas.

CYSTEODEMUS ARMATUS

COMMON NAME Armored blister beetle

DESCRIPTION The most noticeable aspect of this desert-living beetle is its very rotund shape and tough pock-marked yellow elytra. The portly build is no accident, as this ⅜–¾-inch long species traps a ball of air beneath the elytra, which are fused, thus preventing flight. This seems to afford some protection to the body when the beetle is active in the hot sunshine of its arid environment. The all-black *C. wislizeni* is similar.

DISTRIBUTION Deserts of southwestern USA.

MYLABRIS OCULATA

COMMON NAME Eyed blister beetle

DESCRIPTION This is one of numerous brightly colored members of a widespread genus which occurs from southern Africa to Europe and tropical Asia. The adults of this species, which are just under 1 inch long, can be a pest in gardens where they eat flowers, often completely destroying them. The larvae parasitize grasshopper egg-pods beneath the ground.

DISTRIBUTION Southern Africa.

Superfamily **TENEBRIONOIDEA**

DARKLING BEETLES *Family Tenebrionidae*

Somber coloration, especially black, is the norm in this large family with its 17,000 species distributed around the globe. Most of them are small to medium-size beetles which hide away in the day and come out at night to feed, but there are numerous much larger and more attractive desert species which are mainly active in bright sunshine. The larvae develop in all manner of materials.

ELEODES LONGICOLLLIS

COMMON NAME Long-necked darkling or skunk beetle

DESCRIPTION This large ¾–1⅜-inch black beetle can be found scurrying rapidly along over the desert floor from April to September. If closely approached, it instantly stands on its head and points its rear end toward the source of danger. Thus poised, it is ready to blast an attacker in the face with a burst of noxious chemicals. More than 100 closely similar species are found throughout the USA.

DISTRIBUTION Southwestern USA and Mexico.

CTENIOPUS SULPHUREUS

COMMON NAME Sulfur dune beetle

DESCRIPTION This small ¼–⅓-inch long yellow beetle is one of a number of genera which are often included in their own family, the Alleculidae. The larvae live in the ground at the base of plants. The adults are often extremely abundant in July and August on flowers in dry sandy places, such as coastal dunes.

DISTRIBUTION Most of Europe, including the southern part of the British Isles.

Superfamily SCARABAEOIDEA

SCARABS *Family Scarabaeidae*

This is one of the largest, most varied, and most colorful of beetle families, with more than 20,000 species around the world. The antennae bear a distinct club, consisting of several segments which can often be opened out to form a fan. This is particularly noticeable in chafers. True scarabs feed on dung, while chafers and others are herbivores and can become pests.

POLYPHYLLA DECIMLINEATA

COMMON NAME Ten-lined June beetle

DESCRIPTION This large 1–1¼-inch long chafer is easily recognized by the one short and four long white stripes on each elytron. There is a white mark on either side of the head, and three stripes on the pronotum. It is found in woodlands, where the larvae feed on the roots of woody plants. The adults are on the wing in July and August.

DISTRIBUTION Found in the Rocky Mountain areas of Canada and the USA, and in the Southwest regions.

POLYPHYLLA FULLO

COMMON NAME Pine chafer

DESCRIPTION This is the largest European chafer, with an adult length of 1–1½ inches, so therefore unmistakable by size alone, although the marbled pattern is also unique. The males stridulate loudly, and have larger antennae than the female. It occurs mainly near pine woods, where the adults feed on pine needles. The larvae eat the roots of grasses and sedges.

DISTRIBUTION Local in southern and central Europe; absent from the British Isles.

TRICHIUS FASCIATUS

COMMON NAME Bee beetle

DESCRIPTION This is one of the most distinctive and interesting of the European chafers. With its banded pattern, furry exterior, and buzzing flight it mimicks a medium-size bumble bee. The ⅓–½-inch long adults are normally found on flowers, especially thistles, in June and July, chiefly in mountainous areas.

DISTRIBUTION Over most of Europe, but in the British Isles only in the north and west.

TRICHIOTINUS ASSIMILIS

COMMON NAME Flower beetle

DESCRIPTION There are eight species of the genus in the USA and Canada, all in the ⅓–½-inch size range. The adults are rather hairy, and are normally found on flowers, where they eat the pollen. The larvae develop in decaying wood.

DISTRIBUTION North America.

PLUSIOTIS GLORIOSA

COMMON NAME Striped green silversmith or glorious beetle

DESCRIPTION This striking 1–1⅛-inch long species is on the wing in July and August in the western deserts of the USA, often crashing noisily into lighted windows at night. The adults chew the leaves of junipers. Some Central American members of the genus appear to be wrought of gold, hence the name golden beetles.

DISTRIBUTION Texas to Arizona and in northern Mexico.

COTINIS MUTABILIS

COMMON NAME Green fig-eater or green June beetle

DESCRIPTION The larvae of this splendid beetle feed on roots. The 1–1⅛-inch long adults are a shiny semi-metallic green with a brownish margin to the elytra. The adults feed on many kinds of ripe fruit, and are conspicuous by their loud buzzing flight. It is a member of the subfamily Cetoniinae.

DISTRIBUTION Texas to Arizona and adjacent to Mexico.

EUPOECILIA AUSTRALASIAE

COMMON NAME Splendid flower chafer

DESCRIPTION This particularly striking chafer which is nearly ⅞-inch long can be found in daytime on flowers of eucalyptus and other plants, as well as perched on leaves. The beetles chew the petals as well as eating the pollen. The hooked feet typical of chafers can be clearly seen. The larvae develop in decaying wood.

DISTRIBUTION Australia.

MELOLONTHA MELOLONTHA

COMMON NAME Cockchafer

DESCRIPTION The black pronotum of this ⅞–1½-inch long species distinguishes it from the similarly sized *M. hippocastani* and *Anoxia villosa*. All the other mainly brown European chafers are much smaller. Cockchafer adults fly chiefly around dusk, but can be found on leaves and flowers in daytime. The larvae feed on roots and can cause damage to crop and garden plants. The males have much larger antennae than the female.

DISTRIBUTION Common in Europe, including the British Isles.

ORYCTES NASICORNIS

COMMON NAME European rhinoceros beetle

DESCRIPTION With a body length of just over 1½ inches, allied to considerable bulk, this is one of the most impressive European beetles, and is on the wing in June and July. The insect pictured is a male, notable for the long horn on his head; in the female this is just a short point. The larvae are mainly found in old decaying logs. The eastern Hercules beetle *(Dynastes tityus)* from the eastern USA is similar, but gray with brownish flecks.

DISTRIBUTION South and central Europe; absent from the British Isles.

GOLOFA PIZARRO

COMMON NAME Mexican rhinoceros beetle

DESCRIPTION Like all members of this genus, the male beetle pictured has extravagantly developed horns, jutting forward from the head as well as curving upward from its top. The males use these in combination as pincers and levers to gain advantage over a rival during fights over females. Length is just over 2 inches.

DISTRIBUTION Central America.

CANTHON HUMECTUS

COMMON NAME Tumblebug

DESCRIPTION The common name is derived from the way the adult beetles tumble around as they laboriously roll a ball of dung to a nesting site. Both larvae and adults feed on such buried dung balls, often skilfully fashioned from large masses of dung. Sometimes, as in the ½-inch long beetle pictured, a single small, ready-to-use dropping is utilized. In *Canthon* the males and females cooperate to construct and stock the nest. African equivalents are mainly in the genera *Scarabaeus, Sisyphus*, and *Gymnopleurus*.

DISTRIBUTION Mexico; 18 similar species throughout the USA.

STAG BEETLES
Family Lucanidae

One of the most characteristic of all beetle families, the Lucanidae contains some 900 species of mainly large dull-colored insects in which the males have conspicuously elongated antler-like mandibles. In at least one Chilean species these equal the rest of the body in length.

LUCANUS CERVUS

COMMON NAME European stag beetle

DESCRIPTION The male has the large mandibles typical of his sex; those of the females are about one-sixth as large. The males use their mandibles in fights over females, levering one another off a tree-trunk to which the female might come to lay her eggs. Large males measure up to 3 inches long. *L. elaphus* from the northeastern USA is very similar.

DISTRIBUTION Over much of Europe, but decreasing; very local in southern England.

Superfamily **CHRYSOMELOIDEA**

LONGHORN BEETLES *Family Cerambycidae*

Although most common in the tropics, the 20,000 species of this large family are distributed around the world. Their main distinguishing feature is the long antennae, especially in the males, which in some species are many times the body length. The larvae generally develop in wood.

TYPOCERUS ZEBRA

COMMON NAME Zebra flower longhorn

DESCRIPTION This smallish longhorn, which measures about ⅓–½ inch, is very similar to several European *Strangalia* species. Like them, the females can be found laying their eggs in logs, tree trunks, and wooden fence and gate-posts, although the adults are most commonly found on flowers. The larvae develop in dead moist wood. There are 16 *Typocerus* species in North America, several of which differ from this one mainly in the relative amounts of yellow and black.

DISTRIBUTION Forests of eastern USA.

STRANGALIA MACULATA

COMMON NAME Spotted longhorn

DESCRIPTION This is by far the commonest of several similar-looking ½–⅞-inch long European species. Its pattern is very variable, from almost completely yellow to almost completely black. The black and yellow legs and antennae separate it reliably from similar-looking species with all-black legs and/or antennae. It is common on flowers from June through August.

DISTRIBUTION Most of Europe, including the British Isles.

JUDOLIA CERAMBYCIFORMIS

COMMON NAME Tapered longhorn

DESCRIPTION At ⅓–⅖ inch, this is like a small version of the last species, but much more chunky, with the elytra wider and shorter, and the body noticeably tapering from front to back. Large numbers of adults can be found on suitable flowers (e.g. bramble and angelica) from June through August. Several very similar species are found in Europe and across much of North America.

DISTRIBUTION Widespread in Europe, but local in the British Isles.

OCHROSTHES Z-LITTERA

COMMON NAME Z-mark longhorn

DESCRIPTION Open flowery spots in pinewoods in the Mexican high Sierras is the place to look for this small brightly colored and attractive longhorn. Body length is ⅖–⅗ inch and each elytron bears a zigzag yellow mark. The legs are yellow, except the feet which are darker.

DISTRIBUTION Mexico.

CLYTUS ARIETIS

COMMON NAME Wasp beetle

DESCRIPTION This is by far the commonest of the wasp-like longhorns in Europe. Body length is from ¼ inch in males to just over ½ inch in females, thus falling well within the range typical of wasps. The antennae are also unusually short and wasp-like, and the gait jerky. The adults are usually seen on flowers from May through July and also on sticks and dead wood where the females lay their eggs. The North American Douglas fir borer (*C. blaisdelli*) is very similar, but the bands on the elytra are pale whitish-yellow.

DISTRIBUTION Common in Europe, including the British Isles.

MONOCHAMUS OREGONENSIS

COMMON NAME Western sawyer

DESCRIPTION A rather blackish ¾–⅞-inch long beetle spotted with white flecks. Its rather long antennae are ringed with white and are about its body length. It is mainly found in conifer woods, where the larvae develop inside the timber, forming U-shaped tunnels. Several similar-looking species are found throughout North America and Europe.

DISTRIBUTION Western North America.

COSMOSALIA CHRYSOCOMA

COMMON NAME Brown flower longhorn

DESCRIPTION This is one of some 200 similar-looking North American species in several closely related genera. In all cases the tapering body (in females) is in the ⅜–⅞-inch size range; males are smaller. The elytra are normally brown and the long antennae and legs black. The adults are usually found on flowers, especially wild roses. The common and widespread *Stenocorus meridianus* from Europe is very similar in size and color.

DISTRIBUTION Western North America.

TETRAOPES FEMORATUS

COMMON NAME Milkweed longhorn

DESCRIPTION This is one of some 28 similar-looking red and black members of the genus associated with milkweeds. The larvae bore into the stems and roots, while the adults eat the leaves. The eyes are very strange in being divided, with the upper and lower sections widely separated, making in effect four eyes instead of the normal two. The antennae are ringed with white and are quite horizontal to the head.

DISTRIBUTION As a genus found throughout North America.

STERNOTOMIS VARIABILIS

COMMON NAME Variable longhorn

DESCRIPTION There are several similar, spotted, green longhorns found in the forests of tropical Africa. This species, just over 1 inch in length, is particularly common and is usually to be found on the trunks of fallen trees in forests. This female is biting a hole in the bark prior to turning and laying an egg, a common behavior in longhorns.

DISTRIBUTION Widespread in tropical Africa.

TRAGOCEPHALA VARIEGATA

COMMON NAME Variegated longhorn

DESCRIPTION The striking black and orange coloration of this spectacular longhorn is found in only slightly less brilliance in many other members of the genus, which is found in tropical Africa and Madagascar.

This species, which measures just over 1 inch, is found in tropical forests.

DISTRIBUTION Southern Africa.

OBEREA OCULATA

COMMON NAME Willow twig-borer

DESCRIPTION The slim-bodied, approximately ½-inch adults are very difficult to spot when resting on their willow foodplants in July and August. The larvae develop within the twigs. Several more similar-looking species are found in Europe, differing in only minor details from the 25 or so species from North America.

DISTRIBUTION Europe, and the British Isles.

LEAF BEETLES
Family Chrysomelidae

This is one of the largest worldwide families, with over 25,000 species. The adult beetles are normally squarely and chunkily built, often domed and tortoise-like. Almost all the known species feed on leaves. The larvae are soft-bodied and resemble slugs.

CRIOCERIS ASPARAGI

COMMON NAME Asparagus beetle

DESCRIPTION Although only about ¼ inch long, this beetle can occur in such numbers as to inflict severe damage on cultivated asparagus. The black elytra bear four whitish blotches and have red borders. Several similar species are found in Europe, all with rather variable markings. Both larvae and adults of this species feed on asparagus, and the adults can produce quite a loud chirp.

DISTRIBUTION Widespread in Europe (including England) and northern Asia; introduced into North America.

LEPTINOTARSA DECEMLINEATA

COMMON NAME Colorado beetle

DESCRIPTION Still found feeding harmlessly on wild *Solanum* plants in its native home of the southwestern USA and Mexico, this strikingly-marked beetle has become a great pest throughout much of the northern hemisphere on cultivated potatoes. A plague of these ⅜-inch long beetles can ruin the crop.

DISTRIBUTION Spreading in many northern hemisphere countries.

DORYPHORA TESTUDO

COMMON NAME Lurid leaf beetle

DESCRIPTION At about ¾ inch, this striking species is rather larger than any European leaf beetle. It is often abundant on low vegetation where roadways run through areas of rainforest. Its bright yellow and black warning pattern gives notice of extremely noxious properties. There are several similar-looking species.

DISTRIBUTION Over much of central and tropical South America.

CHRYSOMELA POPULI

COMMON NAME Poplar leaf beetle

DESCRIPTION Poplars and willows are the foodplants of this handsome ½-inch long beetle, which is often extremely common where it occurs, for example on coastal dunes. When molested it can release a defensive liquid which reeks of carbolic acid. Several similar-looking members of the genus can do likewise.

DISTRIBUTION Across all of Europe and Asia, to Japan.

DIABROTICA sp.

COMMON NAME Flea beetle

DESCRIPTION Flea beetles are generally rather small insects which leap into the air when disturbed. About ⅕-inch long, this tropical species has been included as it illustrates the enlarged hind femora typical of flea beetles. Many species are serious pests of cultivated plants, e.g. the spotted cucumber beetle (*D. undecimpunctata*), a yellow-bodied species with seven black spots on the elytra. It damages a wide range of crop plants in North America.

DISTRIBUTION The species illustrated is from tropical America.

CASSIDA VIRIDIS

COMMON NAME Green tortoise beetle

DESCRIPTION In tortoise beetles (subfamily Cassidinae) the elytra and pronotum are extended outward to form a flattened tortoise-like carapace, beneath which the legs, head, and antennae can be retracted. This ½-inch long species lives on various mints, especially water mint, and is often found in gardens. There are numerous similar species.

DISTRIBUTION From the British Isles, eastwards to Japan.

METRIONA CLAVATA

COMMON NAME Clavate tortoise beetle

DESCRIPTION This ⅓-inch long species resembles a gall or blotch on a leaf. The edges of the pronotum and elytra are more or less transparent. It is usually found on plants of the morning glory family. Tortoise beetle larvae are strange spiky creatures which hold aloft over their backs a protective shield formed of their molted skins and droppings.

DISTRIBUTION Eastern North America and most of Mexico.

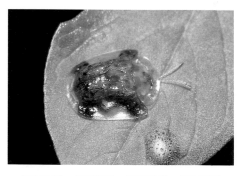

Superfamily CURCULIONOIDEA

LONG BEETLES *Family Brentidae*

The members of this 2,000-strong family are among the most bizarre of insects. The head is extremely elongated so that it is almost of a length with the equally elongated body. The antennae are often beadlike. The majority of species are found in the tropics, especially in areas of rainforest.

BRENTUS ANCHORAGO

COMMON NAME Toothpick weevil

DESCRIPTION This weird ½–1½-inch long creature can often be found in considerable numbers behind the flaking bark of dead and dying trees. This is a male, the females being much smaller and lacking the extended head. Some males are as small as the females. Long males use their snouts as lances to do battle with one another and to guard females as they lay their eggs.

DISTRIBUTION Over most of the tropical Americas; also Florida.

WEEVILS
Family Curculionidae

This is the largest family of insects, with over 40,000 known species. The head is normally prolonged into a snout or rostrum whose tip bears the jaws, the antennae being mounted about halfway along. Many species are wingless, with the elytra fused together. Most species are plant-eaters.

PHYLLOBIUS ARGENTATUS

COMMON NAME Silvered leaf weevil

DESCRIPTION The body of the ⅛–1¼-inch long weevil is covered in glinting metallic greenish scales, although these often wear off over considerable areas in older individuals. The adults are often abundant in May and June on deciduous trees. There are numerous similar-looking species.

DISTRIBUTION Europe (including the British Isles), and much of Asia.

APODERUS CORYLI

COMMON NAME Hazel leaf-rolling weevil

DESCRIPTION This ¼–⅓-inch long weevil
with a rather round body and short
antennae (subfamily Attelabinae) can be
found sitting around on hazel leaves from
May through September. The females
spend many hours painstakingly rolling the
leaves of hazel or birch into cradles which
provide food and lodging for the larvae.
The North American rose weevil
(*Rhynchites bicolor*) is rather similar, but has
a much longer black rostrum.

DISTRIBUTION Much of Europe, including
the British Isles.

CURCULIO NUCUM

COMMON NAME Nut weevil

DESCRIPTION The ¼–⅓-inch body length
of this weevil includes the very elongated
snout. It is found in July and August on
hazel trees, in whose nuts the larvae
develop. The female gnaws into the nut
with her long rostrum before laying an
egg in it. The European acorn weevil (*C.
glandium*) is very similar, as are several
North American species with vastly longer
snouts. In the larger chestnut weevil (*C.
proboscideus*) from eastern North America,
the snout is usually longer than the body.

DISTRIBUTION Europe (including the British
Isles), and much of Asia.

CHRYSOLOPUS SPECTABILIS

COMMON NAME Captain Cook weevil

DESCRIPTION This beautiful ½-inch long weevil is normally black, decorated with a pattern of scintillating green or bluish scales. The adults are usually found sitting around on the leaves of acacia trees in dry forests. The larvae feed on acacia roots and stems, and can sometimes kill young trees.

DISTRIBUTION Australia.

OTIORHYNCHUS SULCATUS

COMMON NAME Vine weevil

DESCRIPTION This rather drab ⅓–⅜-inch long weevil is the scourge of the horticulturalist in many countries. The larvae develop on the roots of many plants and can wreak havoc in nurseries, with container-grown plants being especially susceptible. The adults are rarely seen outdoors, but often noticed walking around on an indoor windowsill.

DISTRIBUTION Almost worldwide.

TALANTHIUM PHALANGIUM

COMMON NAME Daddy-long-legs weevil

DESCRIPTION The long gangly-legged
weevils belonging to the subfamily
Zygopinae are among the most peculiar
members of the family. They run around on
the trunks of trees in forests. The males are
much larger than the females, and stand
over them while they are laying their eggs
in order to fend off rival males. This species
mimics a harvestman (phalangid).

DISTRIBUTION Southeast Asia.

FUNGUS WEEVILS or MOLE BEETLES
Family Anthribidae

The characteristic feature of this family is the way in which the head is extended to form
a short, broad, and rather blunt snout-like structure. Some species are pests of stored
products, particularly coffee; most develop in dead wood. Some 2,000 species are found
around the world.

MECOCERUS RHOMBEUS

COMMON NAME Fungus weevil

DESCRIPTION Most anthribids are rather
drably colored in browns and grays. This
African species, with its pale yellow
marking, is therefore rather striking. Like
many anthribids, it spends its life on tree
trunks, in which the eggs are laid. Note the
broad snout, which in this species bears a
pair of extremely long slender antennae.

DISTRIBUTION Tropical Africa.

SPIDERS

**Over 90 species identified,
with information on body structure,
webs, distribution, and
place in the animal kingdom.**

Introduction

Spiders are everywhere. Every garden or backyard will be full of them, no matter how carefully manicured it may be. Houses are the preferred abode for several kinds of highly urban spiders, while the spider inhabitants of a grassy meadow will be numbered in their millions. Ponds and lakes have their quota and, at the other extreme, deserts are no bar to spiders, which often escape the drought by living in cool burrows deep beneath the sun-baked surface. Even the inhospitable slopes of Mount Everest at an altitude of over 20,000 ft/6,000 m are home to spiders that live nowhere else on Earth.

The number of spider species so far described is just under 40,000, and this is estimated to represent perhaps one quarter of the real total. In Europe, where the tradition of collecting and studying spiders goes back a long way, the total just exceeds 3,000 species. In North America the total is currently rather smaller, but new species are constantly being described.

What is a Spider?

It is quite common to hear spiders being referred to as insects. This is quite wrong. Apart from the fact that insects and spiders are both arthropods, they have little in

Harvestman.

Velvet mite.

Scorpion.

common. Arthropods are animals which have jointed legs and a hard outer skeleton, called an exoskeleton. As the spider or other arthropod grows, it sheds this hard exoskeleton in a series of molts. Insects always have a pair of antennae mounted on the front of the head, their body is divided into three parts (head, thorax, and abdomen), and they have six legs. Spiders never have antennae, they have eight legs, and their body is divided into only two parts: a fused head and thorax called the cephalothorax (or prosoma), and the abdomen (the opisthosoma).

Spiders are members of the class **Arachnida**, which includes some other rather spider-like creatures. The **harvestmen**, belonging to the order Opiliones, generally have long, spindly legs and rather small, rounded bodies. The attachment of the cephalothorax to the abdomen is broad, whereas in spiders there is a narrow stalk. Harvestmen only have two eyes, often perched on turrets, but most

spiders have eight eyes, although some only have six, and a few rare ones only two (but not on "turrets"). **Mites** and **ticks** (order Acari) are mostly very tiny, and it is almost impossible to distinguish the division of the body into head, thorax, and abdomen. **Scorpions** (order Scorpiones) can easily be recognized by their long tails with the stinger at the tip and their large, pincered front legs. **Windscorpions** (order Solifugae) are the sprinters of the arachnid world, running on only six of their eight legs and using the front pair as feelers. **Whipscorpions** (order Uropygi) number fewer than 100 species

Windscorpion.

Whipscorpion, or vinegaroon.

and also walk on just six of their eight legs. The abdomen is clearly divided into segments. They usually have pincer-like pedipalps and long, whiplike tails, although they cannot sting. **Tailless whipscorpions** (order Amblypygi) lack the tail and are more flattened than whipscorpions. The head and thorax are broad, and the abdomen is attached by a stalk. All the legs are slender and held out to the sides, but the front pair is very thin and whiplike, and is not used for walking.

The Spider's Body

As already mentioned, the body is divided into two sections: the **cephalothorax** and the **abdomen**. The top of the cephalothorax is covered by a hard plate called the **carapace**. The front of the carapace bears the group of eyes. These are simple in structure and shine brightly in the light from a flashlight, which is a good way of finding spiders at night. Below the head lie the **jaws**, or **chelicerae**. These are formed of two sections: a stout, fixed basal section tipped by a movable **fang**. In most spiders this can dispense venom through a duct leading from a venom gland. The **mouth** lies below, beneath a lower lip, or **labium**. Spiders dribble copious quantities of digestive juices onto their food (all spiders are carnivores), turning it into a kind of broth. In addition, the basal part of the chelicerae in many spiders is equipped with

Tailless whipscorpion feeding on a cockroach.

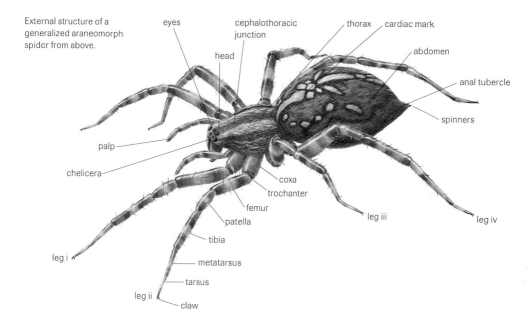

External structure of a generalized araneomorph spider from above.

eyes · cephalothoracic junction · thorax · cardiac mark · abdomen · anal tubercle · spinners · head · palp · chelicera · coxa · trochanter · femur · patella · tibia · leg i · metatarsus · tarsus · leg ii · claw · leg iii · leg iv

teeth that can grind up the prey into an unrecognizable pulp. This is then sucked back up by the spider, using the powerful pumping action of the stomach. Some spiders (e.g. crab spiders, family Thomisidae) pump the digestive juices into their prey and extract the resulting broth from inside, leaving a virtually intact exoskeleton.

Between the jaws and the first pair of legs lie the **pedipalps** (often referred to as the **palps**). These are jointed, and in females and juvenile spiders they are slim and rather resemble small legs. In male spiders the tip of the pedipalps is greatly enlarged and often complex. Before going in search of a female, the male builds a small web on which he deposits a droplet of sperm from the orifice at the rear of the abdomen (**anal tubercle**). He then sucks the sperm back up into the pedipalps. During mating these are then ready for inserting into the female's genital pore (**epigynum**).

The abdomen is relatively soft and flexible. Near its tip lie the **spinners**, which produce various types of silk, depending on its final use. This could be for the external coating of an egg-sac, the wrapping of prey or the spiral strands of a web. Most spiders constantly trail a thin dragline from one of the spinners, and this functions as a kind of safety rope. Silk starts out as a liquid from special glands, but it quickly solidifies as it is stretched. In some spiders (called the cribellates) a flattened, sieve-like plate called a **cribellum** lies just in front of the spinners. By pulling a special row of spines (the **calimastrum**) on the rear leg through the cribellum, very fluffy, multistranded silk, often called **hackled-band silk**, is produced via the spinners.

How To Use The Identifier

The spiders in the identifier section of this book are arranged according to their classification in two groups: the suborders Mygalomorphae and Araneomorphae, which both fall within the order Araneae. A brief description introduces each family, and for each species representing the family the species name and common name are given, together with information on its general characteristics, and known distribution.

When you see a spider, note its surroundings: such as whether or not it is in a web, on the ground, or on a leaf. Consult the information under the family heading and then try to match your specimen to one of the photographs, and the descriptions of the spider's appearance, and habits. This will help you decide whether or not you have chosen the correct genus or species. Lengths given are of the body only and do not include the legs.

Classification of Spiders

Spiders are classified in accordance with certain rules governing the whole of the animal kingdom. This involves an hierarchy, starting with the major grouping, and ending up with the smallest indivisible unit, the species. If we take as an example the garden spider (*Araneus diadematus*), the classification would be as follows.

Kingdom – Animalia
(All animals)
|
Superphylum – Arthropoda
(All animals with an exoskeleton)
|
Phylum – Chelicerata
(Arthropods with unbranched appendages and no antennae)
|
Class – Arachnida
(Terrestrial chelicerates with four pairs of legs)
|
Order – Araneae
(Spiders)

Suborder – Araneomorphae
("True" spiders)
|
Family – Araneidae
(Orb-web spiders)
|
Genus – *Araneus*
|
Species – *diadematus*
(Common name Garden or cross spider)

Suborder – Mygalomorphae
("Primitive" spiders)
|
Family – Theraphosidae
(Tarantulas or bird-eating spiders)
|
Genus – *Brachypelma*
|
Species – *smithi*
(Common name Mexican red-knee tarantula)

The symbols given below accompany each entry and are intended to give vital information about the habits of each spider, and the differences between male and female.

Where is the spider found?

 In open ground such as grasslands, deserts, heaths, and moors.

 In forest and woodland, but often in open spots and along roads.

 Common in gardens.

 Mostly in buildings.

 Mostly in or beside water.

Where is the spider most likely to be sitting?

 In a sheet-web.

 In a silken tube or cell, often under a stone or log, or partly in the ground.

 Inside a burrow.

 On the ground.

 On leaves.

 In an orb-web.

 On the bark of a tree.

On a wall or rock.

 On flowers.

In a tangle- or scaffold-web.

Are the males and females very similar in size and color?

 Very similar.

 A little bit different, mainly in pattern and color, with males being smaller.

Very different. Males are nothing like females, and could be mistaken for a different species.

Suborder Mygalomorphae

The spiders in this suborder are less specialized than those in the suborder Araneomorphae, which are regarded as more advanced. The most significant single difference between the two suborders is the arrangement of the jaws. In the mygalomorphs these have an up-and-down action, moving parallel with the front-to-back axis of the body.

Purse-Web Spiders *Family Atypidae*
This is a small family of stoutly built, shiny-bodied spiders that live in silken tubes. There is a line of sharp teeth along the rim of the basal cheliceral segment, against which the fangs close. This enables the spider to snip holes in the tough silk of the sealed tube and draw prey inside.

ATYPUS AFFINIS

COMMON NAME Purse-web spider
DESCRIPTION The $^{11}/_{16}$ in-/17 mm-long female (illustrated) is a glossy, fat brown spider with very short, stubby legs and huge, projecting jaws. She usually spends her entire life within a closely woven, silken tube resembling the finger of a glove. This tube lies on the ground among grass or under a large stone. When an insect strays onto the silk, it is speared on the spider's long fangs, which stab "blind," but with great accuracy, through the surface of the tube. *Sphodros rufipes* from the eastern USA looks very similar but has redder legs and its buried tube has a section that extends for 8–10 in/20–25 cm up the trunk of an adjacent tree.
DISTRIBUTION Throughout much of Europe southward to North Africa, and eastward to western Asia.

Funnel-Web Tarantulas *Family Dipluridae*

Most members of this small, mainly tropical family can be recognized by the very long spinners. These jut out from the rear of the spider like two tails, and can be more than half as long as the abdomen. These spiders usually construct a broad sheet-web, which can attain 1 yd/1 m in diameter, although some species live in burrows in soft ground.

ANAME sp.

COMMON NAME Australian trap-door spider
DESCRIPTION Despite its common name, this species builds a burrow that lacks a door over the entrance. The 1 in-/25 mm-long male (illustrated) has special spurs towards the tips of his front legs (the left-hand spur is clearly visible in the illustration). These spurs are jammed against the female's fangs during the act of mating, preventing them from operating. At night the males wander in search of females' burrows. If provoked they rear up, open-fanged, in a defensive stance, ready to strike.

DISTRIBUTION Restricted to Western Australia, but similar species occur in other parts of Australia. These include the blackish-brown, stout-bodied Sydney funnel-web spider (*Atrax robustus*), renowned for its deadly bite.

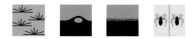

Tarantulas or Bird-Eating Spiders *Family Theraphosidae*
This family includes the largest of all spiders, the goliath tarantula *Theraphosa blondi*, with its 10 in/25 cm legspan. Most tarantulas are very hairy and have a surprising ability to climb on smooth surfaces, aided by dense clusters of hairs on the tips of their legs. The eyes are small and packed closely together. Most species live in burrows in the ground but some live in trees. Prey as large as lizards, mice, and small birds may be taken. Tarantulas are mainly tropical, although some 30 species occur in the USA, with none in Europe.

BRACHYPELMA SMITHI

COMMON NAME Mexican red-knee tarantula
DESCRIPTION This is the most strikingly patterned of all the tarantulas, with its knees boldly marked in bright orange or red. The carapace is black, with a broad, brownish-red margin, and the abdomen is black or dark brown. The female can grow to 3 in/ 75 mm in length, but the male only reaches 2¼ in/56 mm. Like most tarantulas, the adult females are long-lived and spend most of their lives inside deep burrows. The status of this species in the wild has been threatened by the collection of huge numbers of adults for the exotic pet trade, as illustrated.
DISTRIBUTION Mexico.

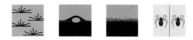

APHONOPELMA CHALCODES

COMMON NAME Western desert tarantula
DESCRIPTION This is one of more than
20 members of this genus found in the
southwestern states of the USA. This
species is often very common in the deserts
of Arizona, and on certain nights large
numbers of males go on the move, searching
for females' burrows. The legs are dark and
the carapace is densely covered with sandy-
colored hairs, while the abdomen is more
sparsely clothed with russet hairs, allowing
a considerable amount of black to show
through. Males reach a length of 1¾ in/
44 mm, females 2¼ in/56 mm.

DISTRIBUTION Arizona, USA, and adjacent
areas of northern Mexico.

HARPACTIRA GIGAS

COMMON NAME Common baboon spider
DESCRIPTION Like most tarantulas, the
baboon spider is seldom encountered
outside its burrow. If threatened, it rears
up, leans backwards on its hind legs, holds
its front legs out at its sides and bares its
fangs, ready to sink them home. However,
the poison is neither potent nor abundant.
The long spinners can be seen projecting
from the rear of the abdomen, while the
carapace bears a spoke-like, radiating
pattern of brown lines on a black
background. Males can reach 2 in/50mm
in length, females ⅓ in/5 mm more.
DISTRIBUTION Cape region of South Africa,
northward to the Transvaal.

Suborder Araneomorphae

The great majority of spiders, both as species and as families, belong in this suborder. The most important difference between these spiders and the mygalomorphs lies in the side-to-side operation of the araneomorph fangs. The members of this suborder are considered to be more advanced and are capable of making many different kinds of silk. The sexual organs are often extremely complex.

Spitting Spiders *Family Scytodidae*
This family of six-eyed spiders can easily be recognized by the hump-backed appearance given by the extreme enlargement of the carapace into a conspicuous dome. This serves to accommodate the enlarged venom glands, which not only manufacture poison but also a special type of glue that is squirted from each fang, forming sticky threads that can immobilize prey up to ¾ in/9 mm away.

SCYTODES THORACICA

COMMON NAME Common spitting spider
DESCRIPTION The eyes of this inconspicuous and rather slow-moving, ¼ in/6 mm-long spider are arranged in three pairs. The domed carapace is obvious from the illustration, as are the yellowish legs ringed with black. The carapace and abdomen are heavily marked with black flecks. As in all spitting spiders, the female (illustrated) carries a ball of eggs in her jaws. There is no web, and the spiders can usually be found under rocks and stones, and in caves and buildings.
DISTRIBUTION Cosmopolitan.

Brown or **Recluse Spiders** *Family Loxoscelidae*
These rather small, brown, six-eyed spiders look harmless enough but are notorious for the
severe symptoms resulting from their bites. They build rather delicate, scrappy sheet-webs
of sticky silk in dark places such as rock crevices, barns, and houses. There are only a few
species worldwide, mostly in the Americas.

LOXOSCELES RUFESCENS

COMMON NAME Violin spider
DESCRIPTION The long legs, held outwards
to the sides, away from the rather slender
brown abdomen, are typical of the group.
The carapace bears a dark mark that has
been likened to a violin, hence the common
name. This ⁵/₁₆ in-/8 mm-long species builds
a small sheet-web under stones, and in
caves and buildings, but also roams actively
in search of prey. It is not aggressive and is
reluctant to bite humans. The brown

recluse (*L. reclusa*) from the USA is much
more venomous and prone to bite, which
can cause severe ulceration. It is larger and
plumper than the violin spider and has a
shinier abdomen.
DISTRIBUTION Native to southern Europe
and North Africa but introduced by man to
a number of areas, including Japan, North
America, Australia, and New Zealand, in
most of which it is thriving and spreading.

Daddy Long-Legs Spiders *Family Pholcidae*

The extremely long, thin legs, with flexible ends, give these spiders a gangling look, which earns them their common name. Most species have eight eyes, among which the front-facing pair at the center are much smaller than the rest. A few have lost a pair of eyes and only have six, in two groups of three. More than 300 species are known worldwide.

PHOLCUS PHALANGIOIDES

COMMON NAME Common daddy long-legs spider or long-bodied cellar spider
DESCRIPTION This ½ in/13 mm-long spider

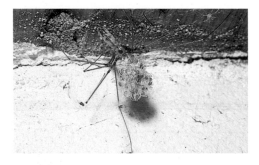

is usually seen hanging motionless for days on end in its large, tangled web in the corner of a room or outhouse. The pale abdomen bears a few dark marks along the top and there is a dark blotch on top of the carapace. The female uses her chelicerae for carrying a ball of eggs, loosely wrapped in a few strands of silk. This species is adept at killing other spiders, including large and powerful house spiders (*Tegenaria* spp.).
DISTRIBUTION Cosmopolitan, mainly in buildings, cellars, and caves.

PHYSOCYCLUS GLOBOSUS

COMMON NAME Short-bodied cellar spider
DESCRIPTION The dark abdomen of this ¼ in-/6 mm-long spider is rather short and plump, and from the side it is triangular in outline. On the face there are eight eyes crowded together on a dark prominence. A blackish band runs up the center of the carapace and is continued onwards for three-quarters of the length of the abdomen. The legs are brown. *P. californicus* from California has a dark, Y-shaped mark on top of the carapace and a pair of dark spots on top of the abdomen.
DISTRIBUTION Throughout the warmer parts

of the world, mainly in buildings, cellars, and caves.

Desert Bush Spiders *Family Diguetidae*
This is a small family of spiders with fewer than 20 species, found in the deserts of the USA, Mexico, and Argentina. They have six eyes in three groups and make characteristic webs in desert shrubs, such as creosote bushes and on prickly-pear cacti. These webs, with their pyramidal-shaped lines, are conspicuous from a considerable distance.

DIGUETIA CANITIES

COMMON NAME Desert bush spider
DESCRIPTION The brownish-orange cephalothorax is rather long and covered in a pelt of very short white hairs. The densely hairy abdomen is light brown and bears on its upper surface a darker, leaf-like shape (folium) bordered with white. The legs are mainly yellowish brown. The easiest way to recognize this ³⁄₈ in-/9 mm-long spider is by its web. This consists of a maze of threads connected to a domed sheet, surmounted by a tubular retreat richly decorated with plant remains and (in females) containing the egg-sacs. The web is often placed some 1–2 ft/ 30–60 cm off the ground in bushes.
DISTRIBUTION Restricted to the USA, from Oklahoma and Texas westward to California, in desert and semidesert.

Six-Eyed Spiders *Family Dysderidae*

This is a small family of spiders having six closely grouped eyes. Rather than two respiratory slits on the underside of the abdomen, there are four, and these are clearly visible. The members of this family are slow-moving, nocturnal hunters with large jaws.

DYSDERA CROCATA

COMMON NAME Woodlouse spider
DESCRIPTION The carapace is a rich reddish brown, while the egg-shaped abdomen is a light tan or gray. The rather thick legs are reddish orange, slightly lighter in tone than the carapace. The fangs are very long and project conspicuously forwards, contributing to an overall length of ⅝ in/16 mm in the females, and ⁷⁄₁₆ in/11 mm in the males. The fangs need to be large in order to penetrate the tough exterior of woodlice, which comprise the main prey of this spider. During the day the spider rests in a flattened silken retreat beneath a stone or log. The bite is painful but not dangerous and is not freely given.

DISTRIBUTION Native to Europe, but now found almost worldwide and common in North America.

Tube-Web Spiders *Family Segestriidae*

Formerly included in the previous family, these spiders now have a small family to themselves. The eyes, six in number, are again closely grouped, and the body is rather tubular, with the front three pairs of legs directed forward for grasping prey. The spiders spend their lives in tubes built in crevices in rocks, walls or tree bark, rushing out at night to seize insects which have blundered across the silken triplines radiating from the tube's entrance.

SEGESTRIA SENOCULATA

COMMON NAME Snake-back spider or leopard spider

DESCRIPTION A row of black spots down the middle of the abdomen, somewhat resembling the pattern of certain snakes, is a distinctive feature of this ³/₈ in-/9 mm-long spider. However, in some specimens (such as the one illustrated) the spots coalesce into an unindented dark band. The dark, shiny brown carapace is rather elongated and the legs are pale brown with a few darker brown rings. The habits of this common species correspond with those given above for the whole family. *Ariadna bicolor*, which occurs throughout the USA, has a similar outline but is a plain dark brown.

DISTRIBUTION Throughout most of Europe, then eastward through Asia to Japan.

Net-Casting Spiders *Family Deinopidae*

The members of this mainly tropical family are also called ogre-faced spiders, on account of two huge, headlight-style eyes staring menacingly from the rather small face (the six other eyes are very small). These huge eyes are essential aids to night vision. Prey is caught in a small net that the spider casts with considerable accuracy. Some species snare their flying insects in mid-air, while others thrust the net downwards to catch crawling insects on the ground beneath.

DEINOPSIS LONGIPES

COMMON NAME Net-casting spider
DESCRIPTION During the day this very slim
⅝ in-/16mm-long brown spider sits
motionless, head-downwards, and is very
difficult to distinguish from a twig. At night
it constructs its prey-catching net of
hackled-band silk. It holds the finished net
in a collapsed state at the tips of its front
legs and then expands it instantly to its full
size while taking aim at a passing insect.
D. spinosa from the southeastern USA is
similar but slightly larger.
DISTRIBUTION Central America.

Feather-Footed Spiders *Family Uloboridae*

More than 200 species of uloborids have been described, mainly from the tropics. One of the main characteristics of this family is the absence of any poison glands. Another peculiarity is the possession of a cribellum, which is used for making an orb-web from hackled-band silk. In most species the front legs bear plumed feet, hence the common name.

MIAGRAMMOPES sp.

COMMON NAME Stick spider (new and undescribed species)
DESCRIPTION Stick spiders build single-line webs of very sticky silk, in which the spider's body forms part of the span. If an insect alights on the convenient perch offered by the single strand of silk, the spider lets go a short length of slack silk so that the line sags and entangles the insect. *M. mexicanus* from Texas is similar.
DISTRIBUTION Indonesia.

Pigmy Mesh-Spinners *Family Dictynidae*

This is the largest family of cribellate spiders, with over 500 species so far described worldwide. Most species are $\frac{1}{10}$ in/3 mm long or less, and the family is mainly distributed in the temperate zones. The small and rather haphazard webs are constructed on leaves, twigs, and in crevices.

NIGMA PUELLA

COMMON NAME Leaf lace-weaver
DESCRIPTION The pale green, rather downy abdomen of the female (illustrated) is strikingly barred with light maroon and there is a large, maroon, rather diamond-shaped blotch near the front. The reddish-brown male is nearly as large as the female. This species lives in a horizontal and rather lacy web built close above the surface of a leaf. More than 150 species of dictynids occur in the USA, most of which are small brown spiders, although some species are quite attractive and are mostly white, yellow or red.
DISTRIBUTION Europe and North Africa.

Large Lace-Weavers *Family Amaurobiidae*
This family of some 350 species contains mainly brown or black spiders with a superficial resemblance to the funnel-weavers in the family Agelenidae. However, the clearly visible cribellum at the tip of the amaurobiid abdomen produces hackled-band silk and serves easily to distinguish the two groups.

AMAUROBIUS FENESTRALIS

COMMON NAME Window lace-weaver
DESCRIPTION The male of this common spider reaches a length of ⁵/₁₆ in/8 mm, while the female (illustrated) can grow up to ³/₈ in/ 9 mm and is considerably fatter. The legs and carapace are a glossy brown and the legs are decorated with a series of darker brown rings. The top of the abdomen bears a dark, wedge-shaped mark girdled by gold. The untidy webs are common around windowframes, under the eaves of houses, and any other places offering a handy crevice, such as the trunks of gnarled old trees. As in several other spiders, the mother's dead body is not wasted but will provide a funeral feast for her offspring.
DISTRIBUTION Europe and western Asia.

AMAUROBIUS FEROX

COMMON NAME Black lace-weaver
DESCRIPTION The plump black ⁵/₈ in-/16 mm-long females are usually found sitting beside their white egg-sacs in a nest underneath a log, rock, or plank of wood. The color of the body is a very dark brown and the abdomen bears a yellowish-brown pattern faintly resembling a human face. The male is slimmer than the female (illustrated) and shorter (⁷/₁₆ in/11 mm).
DISTRIBUTION Europe and North America; recently introduced to New Zealand.

Hole Spiders *Family Filistatidae*
This small family contains fewer than 50 species distributed around the warmer parts of the world. A dozen species occur in the USA. Their most distinctive feature is the rather elongated and pointed carapace, like the prow of a boat, with eight eyes closely grouped together on a small tubercle. There is no epigynum in the female.

KUKULCANIA ARIZONICA

COMMON NAME Arizona black hole spider
DESCRIPTION This black spider has a beautiful, lustrous, velvety sheen. The carapace is distinctly pointed and the male has longer legs than the female (illustrated). During the day the spider rests in a silken tube inside a hole or crevice. A number of silk lines radiate from the mouth of the tube, forming a web, and these are often conspicuous on the walls of buildings. The ½ in-/13 mm-long females live for several years.
DISTRIBUTION Arizona, USA.

Cobweb-Weavers or Comb-Footed Spiders *Family Theridiidae*
This is one of the largest families of spiders, with over 2,200 species worldwide. There are three claws on each leg and most species have eight eyes. The name "comb-footed" stems from the presence of a comb of bristles on the hind legs, which enables the spider to throw silk rapidly, and from a safe distance, over a struggling insect.

THERIDION MELANURUM

COMMON NAME Lesser glasshouse spider
DESCRIPTION As in all members of this family, the spider hangs upside-down in its rather scrappy scaffold-web. The carapace is dark brown, while the abdomen is black and gray. A jagged-edged band down the center of the abdomen is of equal width from front to rear. The light brown legs are dark ringed. The females can usually be found sitting beside their pale brown egg-sacs in the corner of a room or outhouse.
DISTRIBUTION Mainly in houses in Europe and North America.

ACHAEARANEA TEPIDARIORUM

COMMON NAME Glasshouse spider
DESCRIPTION The sides of the globular blackish abdomen are streaked with brown

or gray, while down its center there is a row of pale chevrons. The web is usually constructed in the corner of a room. The outer threads of the web are taut and heavily beaded with glue, which entraps insects much larger than the spider. The pear-shaped egg-sac is suspended in the web, usually with the female on guard beside it. The ¼ in/6 mm long females live more than one year; the males are only ⅛ in/3 mm long.
DISTRIBUTION Cosmopolitan, in houses.

ENOPLOGNATHA OVATA

COMMON NAME Red and white cobweb-weaver
DESCRIPTION The female (illustrated) reaches ¼ in/6 mm in length, and has pale, almost transparent legs and a shiny, brownish-green carapace. The white or cream abdomen can be marked in three different ways: first with a row of dark spots, second with a broad red band, or third with two red stripes. The females are giant-killers, and will fearlessly tackle large stinging insects, such as the bumble-bee in the illustration. The female stands guard over her grayish-blue egg-sac inside a curled leaf fastened together with silk.
DISTRIBUTION Native to Europe, but introduced to North America where it is

now a common and familiar spider.

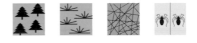

ARGYRODES FLAVESCENS

COMMON NAME Red and silver dewdrop spider
DESCRIPTION The reddish-brown, very shiny abdomen bears a number of silver spots and, as in most dewdrop spiders, it is high and domed, with an almost triangular outline. Most members of this genus are tiny (⅛ in/3 mm long or so) and live in the webs of other larger spiders (usually orb-web spiders such as *Nephila* and *Argiope*), feeding on the prey caught by the host spider. *A. trigonum*, a common species in the USA, looks rather similar, but is more yellowish and lacks the silver spots.
DISTRIBUTION Sri Lanka and Burma eastward through Indonesia.

STEATODA GROSSA

COMMON NAME Cellar spider
DESCRIPTION The male (illustrated) reaches
¼ in/6 mm in length. The legs are a pale,
almost transparent brown, the carapace
a glossy, deep brown (almost black), and
the abdomen black with several greenish-
brown marks, on which there is a peppering
of white spots. The larger female (up to
⅝ in/16 mm long) has a much rounder,
glossy black abdomen, with no visible
markings (except sometimes a pale
semicircle and three spots), and a black
carapace and legs. The female suspends
her fluffy white egg-sacs in her web, which
is usually found in cellars, caves or hollow
trees. She guards her egg-sacs until the
babies emerge, after which they remain

in her web for some time, sharing her food.
DISTRIBUTION Cosmopolitan.

LATRODECTUS MACTANS

COMMON NAME Black widow spider
DESCRIPTION The red "hourglass" marking
on the underside of the shiny black abdomen
is the unmistakable trademark of the black
widow spider. The round-bodied females

reach a length of ⁹⁄₁₆ in/14 mm but the much
paler males only reach ³⁄₁₆ in/5 mm. The
very scrappy web is often found in and
around buildings – the web in the illustration
was underneath a table in a house. The bite
is highly venomous, but seldom lethal in
man, and the black widow is a nervous
spider which will drop from its web at the
slightest disturbance. In the almost identical
species, *L. hesperus*, enemies such as mice
are repelled by having their faces smeared
with sticky silk that the spider deploys in
a kind of defensive net.
DISTRIBUTION Many of the warmer parts of
the world, including southern USA.

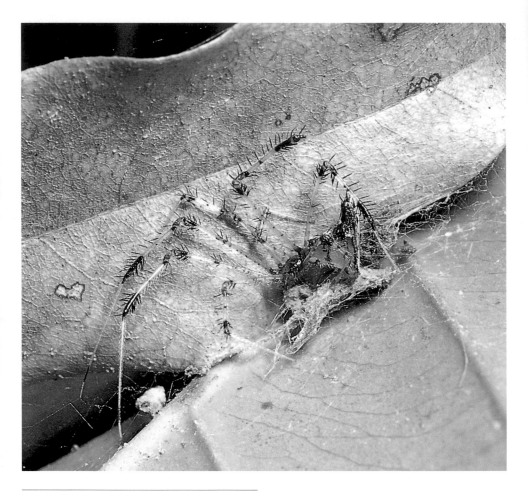

CHRYSSO sp.

COMMON NAME Triangular spine-leg spider
DESCRIPTION This is a very strange-looking spider. The high-peaked abdomen is triangular in profile and very flattened. The unusually long legs are fringed with numerous, stiff black spines. The untidy web is hung beneath the foliage of trees and bushes, and the ⁵/₁₆ in-/8 mm-long female hangs upside-down, often beside her tan-colored, bell-shaped egg-sac. The species illustrated is presently undescribed. *C. pulcherrima* is a cosmotropical species which is similar but less spiny.
DISTRIBUTION Indonesia.

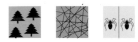

Dwarf Spiders or Money Spiders *Family Linyphiidae*

This is a large family of mostly very small spiders which are more common in the cooler parts of the world, such as Europe and North America, than in the tropics. The males of many species are provided with stridulatory ridges on the chelicerae which are used to produce vibrations of the web during courtship. In addition, there are many instances where the head of the male is embellished with bizarre knobs and turrets which play a part in the mating process. The web is usually a small hammock. Dwarf spiders are the most abundant of all spiders and their populations can reach 2¼ million per acre/5½ million per hectare in a suitable habitat, such as a grassy field.

NERIENE PELTATA

COMMON NAME Platform-web spider

DESCRIPTION This is one of several species of black (or brown) and white dwarf spiders which can be seen hanging beneath their hammock-webs during the day. The abdomen of this ³⁄₁₆ in-/5 mm-long species bears a narrow white central stripe, bordered on either side by a wavy-edged brown band, flanked by white. The carapace is light brown with a triangular blackish central band. The web is mainly built in bushes and among the lower branches of trees.

DISTRIBUTION Europe, eastward to Japan, and in the USA.

DRAPETISCA SOCIALIS

COMMON NAME Invisible spider
DESCRIPTION This ⅛ in-/3 mm-long spider
spends the day sitting motionless on the
trunk of a tree. The mottled, brown and
white abdomen, and banded brown and
black legs blend in very well with the colors
and textures of tree bark. Usually the spider
is perched on top of a fine web that hugs
the bark. *D. alteranda* from the USA is very
similar and occurs on tree trunks from
New England westward to Wisconsin.
DISTRIBUTION Throughout Europe and most
of temperate Asia.

MICROLINYPHIA PUSILLA

COMMON NAME Dainty platform spider
DESCRIPTION The shiny black male
(illustrated) reaches a length of ³⁄₁₆ in/5 mm
and has a tubular abdomen with two white
spots on the top, near the front. The slightly
larger female is completely different and
looks more like *Neriene peltata* (page 31),
with a plump silvery abdomen bearing a
broad, black, leaf-like mark. The small
hammock-web is built among grass and in
low vegetation, often in damp places.
DISTRIBUTION Europe, temperate Asia, and
North America.

Funnel-Weavers or Grass Spiders *Family Agelenidae*

This is a large family containing more than 1,000 species of mainly brownish spiders, varying in length from ¹⁄₁₆ in/2 mm to 1 in/25 mm, and usually having long and conspicuously hairy legs. They construct a broad, flat sheet-web of non-sticky silk leading down into a tubular retreat. Intersecting knock-down lines deployed above the sheet bring flying insects crashing down into the web.

AGELENOPSIS APERTA

COMMON NAME Desert grass spider

DESCRIPTION There are numerous species of grass spiders found throughout the USA but these can only be reliably distinguished from one another by an expert. The males average ⁵⁄₈ in/16 mm in length, the females ³⁄₄ in/19 mm. There are eight eyes, arranged in two rows. Most species have a pair of dark stripes on either side of the carapace. The spider illustrated is one of the grayer species; most of the others are brown. There is usually some kind of pale, longitudinal striping on the abdomen, from which the spinners protrude conspicuously. The spider lurks at the entrance to its tubular retreat, ready to rush out and pounce on arriving insects. The females lay eggs in autumn, then die. *Agelena labyrinthica*, from Europe, is similar but brown.

DISTRIBUTION Numerous similar species are found throughout North America.

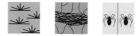

TEGENARIA DUELLICA

COMMON NAME Cobweb spider
DESCRIPTION This is the spider that is most likely to cause panic as it sprints rapidly across the carpet on its long, hairy legs, or crouches menacingly in the bathtub. It is, of course, perfectly harmless but is responsible for eliciting fear and loathing in countless

people. The ½ in-/13 mm-long males are often the culprits, as they abandon their large sheet-webs in the corner of a room or outhouse and wander off in search of females. These Romeos usually move in with the slightly larger females and they will live together amicably for a considerable time before the male finally dies. His corpse provides his mate with a substantial meal that will make a significant contribution towards the development of her eggs, in effect converting father into father's offspring. This spider, along with several of the other large species of *Tegenaria*, can endure long periods of fasting and the females may live for several years.
DISTRIBUTION Europe and North America.

TEGENARIA AGRESTIS

COMMON NAME Yard spider
DESCRIPTION This European spider has been introduced to the northwestern states of the USA, where it has become common around houses. Its large, untidy-looking sheet-web is usually found in backyards. It is smaller and rather grayer than the cobweb spider, the female (illustrated) reaching a length of $^9/_{16}$ in/14 mm but the male only $^3/_8$ in/9 mm. The spherical egg-sacs are placed in the web structure and are camouflaged with bits of rotting wood, particles of soil or insect skeletons. Numerous instances of this species biting humans have been reported but the bite is not serious.

DISTRIBUTION Europe and northwestern North America.

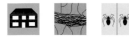

Long-Jawed Orb-Weavers *Family Tetragnathidae*

The orb-webs constructed by these usually elongated spiders are seldom vertical but are usually inclined at an angle and sometimes horizontal. With the exception of *Leucauge* the females lack an epigynum. The jaws in both sexes are large and prominent, and the males of some species have special prongs for jamming open the female's jaws during mating. Members of the genus *Pachygnatha* have more rotund bodies and do not construct webs as adults. Within this family are spiders of the genus *Nephila*, the largest of all web-building spiders. The incredibly strong yellow webs of *Nephila clavipes* in the USA have been recorded spanning gaps as much as 60 feet across.

TETRAGNATHA EXTENSA

COMMON NAME Common long-jawed orb-weaver

DESCRIPTION The elongated body and legs are typical for the genus. This ⅜ in-/9 mm-long species is most likely to be found near water. During the day the spider sits in the center of the web or on a nearby leaf or grass stem, with the long front and back legs held out fore and aft. They often lie hidden from their predators and in wait for possible prey items. The egg-sacs are placed on a leaf and are covered in tufts of grayish-green silk. These spiders build impressive webs that are capable of trapping larger warm-blooded animals, such as humming birds and bats, although the normal prey is much smaller.

DISTRIBUTION Europe and most of temperate Asia, and in North America.

METELLINA SEGMENTATA

COMMON NAME Common orb-weaver
DESCRIPTION This is probably the commonest
orb-weaver in Europe. The ⁵/₁₆ in-/8 mm-
long female (illustrated) is pale yellowish
brown and there is a darker pattern
resembling a leaf or fir tree on the abdomen.
The male has longer legs and is a more rusty
brown, with a smaller and slimmer abdomen.
During the mating season the males take up
residence in the female's web but wait until
an insect is trapped before commencing
courtship. They have to entice the female
on to a special mating thread before
consummation can take place.
DISTRIBUTION Europe, eastward to Asia,
and in Canada.

LEUCAUGE NIGROVITTATA

COMMON NAME Black-striped orchard spider
DESCRIPTION Orchard spiders can be easily
recognized wherever they are seen
throughout the world, as they all have a very
similar and highly characteristic appearance.
This ¹/₂ in-/13 mm-long species has a silver
stripe down the center of the abdomen,
dissected lengthwise by three black lines;
the flanks are greenish yellow. The webs are
inclined at 45° to the vertical and are often
grouped together.

Two species of orchard spiders occur in
the eastern USA. In the Mabel orchard
spider (*L. mabelae*) the yellowish abdomen
bears eight, evenly spaced silver bands, and
three orange spots. In the venusta orchard
spider (*L. venusta*) the top of the abdomen
is silver, striped with black, the sides are
yellow, and there are two red spots on the
underside. Both species build an almost
horizontal web in the lower branches of
shrubs and trees.
DISTRIBUTION Indonesia.

NEPHILA CLAVIPES

COMMON NAME Golden-silk spider or golden orb-weaver
DESCRIPTION The females are the giants among orb-weavers and can reach 1⅜ in/ 34 mm in length. The front two pairs of legs, along with the rearmost pair, all bear tufts of dark hairs. The sausage-like abdomen is decorated with white or gold spots and flecks, while the carapace is silver. The brown males are only ⅜ in/9 mm long. The huge webs are made of tough yellow silk, and often carry uninvited dewdrop spiders (*Argyrodes* spp.) klepto-parasites. In the USA the adults die in winter but in the tropics they are present throughout the year.

DISTRIBUTION Southeastern states of the USA, southward to northern Argentina.

NEPHILA SENEGALENSIS

COMMON NAME African golden orb-weaver
DESCRIPTION The top of the female's abdomen has a broad black or silver gray, rather jagged-edged band in which are set four pairs of pale cream spots. These are often united to form discrete bands or else form a dumb-bell shape. The sides of the abdomen are also cream, and the carapace is blackish silver. The legs are banded in black and dark brown. The juveniles are much more silvery, with numerous whitish spots, and are often seen sitting below a dense stabilimentum of thicker silk, adorned with numerous insect remains. Adult females may be 1⅜ in/34 mm in length, but the brown males only reach ¼ in/6 mm and are usually present in the females' webs.

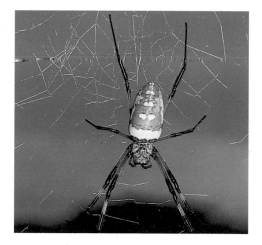

DISTRIBUTION Most of sub-Saharan Africa, in desert, savannah, and forest.

Orb-Weavers *Family Araneidae*

This is a large and varied family, with over 3,500 species around the world. Although the perfect orb-web is perhaps a typical feature of the family, there are huge variations in its structure, and some species have dispensed with a web entirely and reverted to a life of ambush. Orb-weavers have tiny eyes that play only a minor role in daily life. The sense of touch is far more important, and these spiders monitor events in their webs through their vibration-sensitive legs and feet.

ARANEUS DIADEMATUS

COMMON NAME Garden or cross spider
DESCRIPTION The color is very variable and can be pale fawn, deep rusty brown, bright orange, brownish-gray and every tone in between. Whatever the ground color, in the female (illustrated) there is usually an unmistakable cross on top of the abdomen, formed by a series of white blotches. The spider sits in the center of its web or on a plant nearby. Most prey is bitten first, then wrapped. Females attain a length of ⅝ in/

16 mm and become very plump when they are full of eggs; the slimmer male does not exceed ⅜ in/9 mm.

DISTRIBUTION Europe, eastward to Japan. Introduced to North America and now very common in many areas, especially in the east.

ARANEUS QUADRATUS

COMMON NAME Four-spot orb-weaver
DESCRIPTION The female has a rotund abdomen which is usually tan colored but can be rusty brown or greenish yellow. There are always four prominent white spots on the front half of the abdomen, plus a number of small white dots and squiggles. The legs are attractively banded. The spider rarely sits in its web, but can usually be located sitting head-downward in a lair made by fastening together a number of plant stems with a dense mesh of silk. The shamrock spider (*A. trifolium*), common throughout North America, more or less combines the patterns of the four-spot orb-weaver and the garden spider (*A. diadematus*), with the four spots of the former and the vertical section of the cross of the latter.

However, it has a white carapace with a central black stripe, and another black stripe on each side, a distinctive character absent in the other two species.
DISTRIBUTION Throughout Europe and much of Asia.

ARANEUS ILLAUDATUS

COMMON NAME Texas orb-weaver
DESCRIPTION This is a very large, grayish-white or sometimes even pinkish-white species in which the female (illustrated) can reach 1 in/25 mm in length. The top of the abdomen (near the front) bears two blackish-brown markings, which are more or less triangular and have very indented, rather ragged edges. Each of these darker markings contains a white dot, and there are some more, rather leaf-like markings towards the rear of the abdomen. The whole spider is extremely hairy. The male is a pigmy of only ³/₈ in/9 mm.
DISTRIBUTION Southeastern Arizona to western Texas.

ARANEUS MARMOREUS

COMMON NAME Marbled orb-weaver
DESCRIPTION There are two forms of this species. In the one illustrated (var. *pyramidatus*) the female's plump abdomen varies from pale whitish-cream to a rich golden yellow. Towards the rear there is a dark brown, pyramid-shaped blotch. In the other form the abdomen is a much darker shade of yellowish orange and bears a definite brown or blackish pattern of markings with zigzag edges, giving a marbled effect. The two forms are seldom found together and var. *pyramidatus* only occurs in the European populations. The female reaches a length of $^9/_{16}$ in/14 mm and the male $^3/_8$ in/9 mm. The web is placed among low vegetation in open areas in woodland, and on heaths.
DISTRIBUTION Europe, eastward over most of Asia, and in most of North America but absent from the southwestern USA.

ARGIOPE APPENSA

COMMON NAME Great golden argiope
DESCRIPTION The female (illustrated) can often reach a length of 1¼ in/32 mm. The top of the abdomen is a bright lemon yellow, with a pair of large black dots, and two smaller ones. The legs are strongly banded in black and gray. The webs are often built in gardens, or even under the eaves of houses, where several may occur together. **DISTRIBUTION** Southeast Asia.

ARGIOPE ARGENTATA

COMMON NAME Silver argiope
DESCRIPTION In this handsome spider the carapace and the front half of the abdomen are bright silver. The rear half of the abdomen is brightly decorated with silver markings on a yellow or orange background, and there are five prominent lobes around the rear margin. The shape of the abdomen is very characteristic, as it is narrow at the front and then flares out towards the rear. The legs are banded in black and yellow.

The female (illustrated) can reach a length of ¹¹/₁₆ in/17 mm, but as in all *Argiope* spp., the male is a midget and only reaches ³/₁₆ in/5 mm. He is also much more lightly built than the female. The web is built near the ground among grasses and shrubs.
DISTRIBUTION From southern USA (southern Florida westward to southern California) to northern Argentina.

CYRTOPHORA HIRTA

COMMON NAME Hairy tent-spider
DESCRIPTION The ground color of the ⁹/₁₆ in-/14 mm-long female (illustrated) is silvery white. The top of the abdomen is decorated with a complex black pattern, and there is a yellow and black pattern on its sides. The spinners are situated on a tubercle, which gives the rear of the abdomen a distinctively prow-shaped profile from the side. The legs are banded light and dark. The spider hangs upside-down beneath a horizontal orb-web, which is slung beneath a scaffold of intersecting lines. These interrupt flying insects and knock them down into the web. None of the silk is sticky, which is rare for an araneid. Several webs occur together.
DISTRIBUTION Australia.

ZYGIELLA X-NOTATA

COMMON NAME Missing sector orb-weaver
DESCRIPTION The webs of this common spider are often abundant and conspicuous on and around houses, especially on window frames, and beneath roof overhangs. The webs are easily recognized by the absence of one section of the orb, near the top. During the day the spider hides in a silken lair nearby. The smooth, silvery abdomen is marked with a darker leaf-like pattern and the legs are heavily banded. Females reach a length of ⁵/₁₆ in/8 mm while the males are slightly smaller but look very similar.

DISTRIBUTION Common in Europe, much of temperate Asia, and the whole of North America.

CYCLOSA CONICA

COMMON NAME Conical orb-weaver
DESCRIPTION The color of the abdomen varies from gray with black markings to almost pure black. The abdomen is distinctly, but bluntly pointed, at the rear end and slopes forward over the carapace at the front. The females can reach ¹/₅ in/5 mm in length; the males are slightly smaller. The spider usually sits at the hub of its surprisingly large web. Above and below the hub there is usually a vertical string of prey remains incorporated into a dense silk stabilimentum. In *C. turbinata* from the USA the tip of the abdomen is extended into a longer and more acute point.
DISTRIBUTION Europe, Asia, and North America.

GASTERACANTHA ARCUATA

COMMON NAME Long-horned orb-weaver
DESCRIPTION The ³/₈ in-/9 mm-long female
(illustrated) of this amazing spider can be
mistaken for no other. From the top corners
of the almost triangular, bright orange
abdomen protrude two long, black, curved
"horns." Two thinner and shorter spikes jut
up from the rear margin of the abdomen,
between the major horns. It is believed that
the projections make it difficult for birds
to grasp hold of them, though no one is
absolutely sure. The legs are black and the
rather inconspicuous carapace is dark
brown. The tiny males are hornless.
DISTRIBUTION Asia, from India to Indonesia,
often in gardens.

GASTERACANTHA HASSELTI

COMMON NAME Hasselt's spiny spider
DESCRIPTION In the $^5/_{16}$ in-/8 mm-long female (illustrated) the top of the almost triangular abdomen is glossy bright orange, with a diverging row of six black spots on either side of the center line. The top corners are drawn out into two thick black spikes, which are more or less straight and have pointed tips. Two smaller black spikes protrude from the rear margin of the abdomen and two more from the sides, midway between the large spikes and the carapace, which is densely covered in short white hairs.
DISTRIBUTION India to Indonesia.

GASTERACANTHA STURII

COMMON NAME Blunt-spined kite spider
DESCRIPTION The very broad, kite-shaped abdomen is glossy yellow, crossed by two black lines. The top corners of the abdomen are armed with two stout, black, and rather hairy spines. These are blunt-ended save for a tiny thorn at the tips. Two very short spines project from the front part of the abdomen and two more from the rear. This stur's female kite spider sits on a leaf in the rain forest margin.
DISTRIBUTION Southeast Asia.

GASTERACANTHA FALCICORNIS

COMMON NAME Horned orb-weaver
DESCRIPTION The very hard, bright red, glossy abdomen of the female (illustrated) is deeply punctured with black pits. Two long curved horns and four short straight ones project from the abdomen. Sometimes a white or yellow band runs across the abdomen. This is quite a good example of a typical araneid orb-web although it is damaged with use.
DISTRIBUTION Eastern and southern Africa.

GASTERACANTHA CANCRIFORMIS

COMMON NAME Crablike spiny orb-weaver
DESCRIPTION The ½ in-/13 mm-wide abdomen of the female (illustrated) is subject to considerable color variation. The ground color can be white, cream, yellow or pale orange, and the six spurs can be black or red. Some forms are liberally speckled with black or brown spots, while others are virtually unmarked. The carapace is dark brown. The spider normally hangs from the underside of a sloping web, built near ground level. This female was found in a Mexican rain forest.
DISTRIBUTION USA, from North Carolina to Florida, and across to California; Central America.

GASTERACANTHA MINAX

COMMON NAME Christmas spider
DESCRIPTION In the female (illustrated) the ground color of the abdomen is black, with a white or yellow pattern. Six stumpy spurs circle the abdomen, giving it a star-shaped appearance. The width of the abdomen is about $^7/_{16}$ in/11 mm. The webs are constructed on shrubs and often occur in vast numbers over hundreds of acres.
DISTRIBUTION Over most of Australia.

MICRATHENA GRACILIS

COMMON NAME Lumpy thorn spider
DESCRIPTION The color of the abdomen can be white, pale yellow, yellowish brown or even black, while it can be virtually unmarked or spotted with brown. The $^3/_8$ in-/9 mm-long female (illustrated) has ten sharp spines. The spinners are set well forward on a deep tubercle, so that in profile the abdomen is almost triangular. The shiny carapace is elongate and the eyes are tiny. The male's spineless abdomen is long and almost parallel sided.
DISTRIBUTION USA, east of the Rockies; Central America.

MICRATHENA SAGITTATA

COMMON NAME Arrow-shaped thorn spider
DESCRIPTION The top of the abdomen is bright yellow, with a number of black puncture marks. Two long, deep-red spines protrude from the top corners of the abdomen and there are two pairs of shorter spines at even intervals along the sides. The yellow underside (illustrated, in a female) is patterned with red and black, and the spinners protrude downward on a distinct tubercle. Females are ³/₈ in/9 mm long, males only ³/₁₆ in/5 mm.
DISTRIBUTION Eastern USA and Central America.

MECYNOGEA LEMNISCATA

COMMON NAME Basilica spider
DESCRIPTION In its shape and color this ³/₈ in-/9 mm-long spider is rather similar to the *Leucauge* spp. orchard spiders. However, in the basilica spider the abdomen has a conspicuous hump on either side, near the base. The yellow carapace has a narrow black line down the middle and a wider line along each margin. The olive-green abdomen sports a black and brown, leaf-like pattern, bordered with white, and there is a wavy white line down each side. The snare consists of an orb-web arranged as a horizontal dome, set between upper and lower scaffolds of intersecting lines. The spider sits head-downward and, in the fall, the females will be hanging beneath a string of egg-sacs, as seen in the illustration.
DISTRIBUTION Most of the USA, but excluding the west-coast states.

Wolf Spiders *Family Lycosidae*

This is a large family of mainly brown or gray spiders, with more than 3,000 species worldwide. The majority are active hunters, or sit on leaves and pounce on insects that walk or fly past. Most wolf spiders live on the ground, and many do not have any kind of permanent home, although others live in silk-lined burrows, emerging after dark to hunt. Their vision is good, aided by two large eyes in the center of the face, flanked by two smaller eyes. A further row of four small eyes lies beneath them.

TROCHOSA TERRICOLA

COMMON NAME Earth-chaser
DESCRIPTION The legs are light brown, and the abdomen and carapace are a slightly darker brown, with few obvious markings. The ½ in-/13 mm-long female (illustrated) spends the day in a silken cell beneath a stone or log, emerging at night to hunt.
DISTRIBUTION Europe, much of Asia, and northern USA.

PARDOSA AMENTATA

COMMON NAME Spotted wolf spider
DESCRIPTION Dozens of similar-looking species of *Pardosa* wolf spiders are found in all kinds of habitats in Europe and North America (more than 50 species in the USA). As in all of them, the $^5/_{16}$ in-/8 mm-long female of this species carries her egg-sac attached to her spinners. The babies ride around on their mother's back for the first few days. The males are only slightly smaller than the females and employ a system of semaphore-signaling with the legs and palps during courtship.

DISTRIBUTION Europe and much of temperate Asia.

LYCOSA CAROLINENSIS

COMMON NAME Carolina wolf spider
DESCRIPTION This is the largest of a multitude of wolf spiders found in the USA. The females can reach a length of $1^3/_8$ in/34 mm, the males $^3/_4$ in/19 mm. The carapace is rather broad and can be quite dark, as in the specimen illustrated, with only a narrow pale central band. In some individuals the central band is broader and lighter and there are similar bands along both sides of the carapace. The abdomen varies from light brown to very dark brown, usually with a blackish-brown central band with either straight or jagged margins. The females spend much of their lives in a burrow in the ground but emerge at night to hunt, often carrying their large white egg-sacs attached to the spinners.
DISTRIBUTION Throughout the USA and much of Canada.

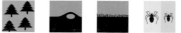

PIRATA PIRATICUS

COMMON NAME Common pirate spider
DESCRIPTION Members of this genus can be recognized by the pale V- or Y-shaped mark on top of the carapace. There are numerous similar-looking species that live beside freshwater in Europe and the USA. This $5/16$ in-/8 mm-long species has a light brown, velvety abdomen with a distinct sheen. A pair of pale lines runs from the front of the abdomen to the rear, gradually converging so that they eventually touch to form an elongated V. There is a line of small white spots toward the rear of the abdomen, whose sides are densely clothed in short, white hairs. The female (illustrated) carries her spherical white egg-sac attached to her spinners and runs with it across the surface of the water. If alarmed, she will disappear beneath the surface for several minutes.
DISTRIBUTION Europe, eastward through Asia to Japan, and in North America.

ARCTOSA PERITA

COMMON NAME Sand-runner
DESCRIPTION The members of this genus are characterized by a rather flattened carapace, with the posterior eyes more or less mounted on top so that they look upwards. The carapace of the female (illustrated) varies from pale brown to black, the abdomen bears four large pale spots against a darker background, and the legs are ringed dark and light. This generally mottled effect gives excellent camouflage against a sandy background. The spiders spend part of their time in a burrow in the sand but also emerge to hunt in daytime. There are around a dozen *Arctosa* spp. in the USA, of which *A. littoralis* is very similar to the sand-runner, being found throughout the USA.

DISTRIBUTION Europe and much of Asia.

Nursery-Web Spiders *Family Pisauridae*
These are similar to the wolf spiders but the eyes are more or less all the same size.
The female also carries her egg-sac, as in wolf spiders, but holds it beneath the front of her
body, suspended from her fangs and pedipalps. The females make silken nurseries for their
young. About 500 species occur worldwide.

PISAURA MIRABILIS

COMMON NAME Wedding-present spider
DESCRIPTION Males and females are of
approximately equal size, averaging ⁹/₁₆ in
/14 mm in length, but the male is of a slightly
lighter build and has longer front legs. The
ground color varies greatly and can be gray,
light tan, chocolate brown or anything in
between. Down the center of the carapace
there is a white line, flanked by a dark band.
The sides of the abdomen usually bear a
dark, wavy-edged band, while the top usually
has a series of faint chevrons. However, in
some specimens all markings are virtually
absent. The unique feature of this spider is
the male's habit of presenting the female
with a "wedding present" consisting of a fly
densely wrapped in white silk. The female
can often be seen trundling around with her
large, spherical white egg-sac. Just before
the young emerge, she fastens a few leaves
or stems together with silk, places the egg-
sac within, and then stands on guard. The
American nursery-web spider (*Pisaurina
mira*) looks similar, is the same size, and has
similar nesting habits but does not present
a nuptial gift. It is common throughout
most of the USA east of the Rockies.
DISTRIBUTION Europe, eastward through
most of temperate Asia, and in North Africa.

DOLOMEDES FIMBRIATUS

COMMON NAME European fishing spider
DESCRIPTION This dark brown spider is easily recognized by the white band that runs along both sides of the body. The female (illustrated) can reach a length of ³/₄ in/19 mm, the male only ¹/₂ in/13 mm. They sit on water, waiting to detect the ripples from an insect or fish. The females carry their large, grayish-green egg-sacs in their chelicerae. The six-spotted fishing spider (*D. triton*) from the USA looks very similar and has the same habits. It is common in swamps east of the Rockies.
DISTRIBUTION Europe and much of Asia.

Lynx Spiders *Family Oxyopidae*

These agile spiders can be recognized by the slim, tapering abdomen which ends in a point, long, heavily spined legs, and unique eye arrangement, with six larger eyes forming a hexagon and two smaller eyes below. The females guard their egg-sacs. Most species are found in the tropics, with only some 15 in the USA and even fewer in Europe.

OXYOPES SCHENKELI

COMMON NAME Bridal-veil lynx spider
DESCRIPTION The pattern of this ³⁄₈ in-/9 mm-long spider is so variable that it is impossible to describe in words, so perhaps likening it to a Persian carpet will adequately sum it up. It can leap into the air to capture a passing butterfly. Prior to mating, the female hangs in mid-air, suspended from her dragline, while the male wraps her in a silken bridal veil. This is the only lynx spider known to exhibit such behavior.

DISTRIBUTION Tropical Africa.

PEUCETIA VIRIDANS

COMMON NAME American green lynx spider
DESCRIPTION The normal color of this beautiful ³⁄₄ in-/19 mm-long spider is bright green but some individuals, especially in the western states of the USA, are yellowish or brownish. The top of the abdomen is usually decorated with a double row of red chevrons but these may be very faint. The closely spaced eyes are situated within a pale brown rectangle and the legs are very spiny. The females fix their knobbly, straw-colored egg-sacs to a leaf and then stand guard over them until the young hatch. This spider has the amazing ability to "spit" venom into the eyes of an aggressor. Similar species are found in southern Europe, and Africa.
DISTRIBUTION USA, from New England southward to Georgia and westward to the Rocky Mountains, and in Central America.

Sac Spiders *Family Clubionidae*

This is a family of more than 1,500 species of mainly brown nocturnal spiders. They spend the day in a silken tube built inside a rolled leaf or placed beneath a stone or log. They are active hunters and do not build prey-catching webs. Many species are superb mimics of ants.

CHEIRACANTHIUM ERRATICUM

COMMON NAME Grass-head sac spider
DESCRIPTION The legs are a translucent yellowish brown. The carapace is brown, and the velvety abdomen straw-colored, with a broad yellow central band bisected by a maroon median stripe. The females enclose themselves and their eggs in a nest, which is usually composed of a grass-head fastened together with silk to form a disk. *C. mildei*, found in southern Europe and North America (introduced) is very similar but is often found in buildings, where it has bitten humans, resulting in nasty skin-blisters.
DISTRIBUTION Europe and much of Asia.

CASTIANEIRA spp.

COMMON NAME Painted ground spider
DESCRIPTION More than 25 species of *Castianeira* are found in the USA. Many of these mimic ants, or are brightly colored and mimic velvet ants (mutillid wasps). Most species raise and lower their abdomen and front legs as they scuttle around on the ground, rather in the manner of ants or wasps. The egg-sacs are in the form of a flattened disk with a metallic sheen and are usually fastened to the underside of a rock. The species illustrated is ³/₈ in/9 mm long and was in the Arizona desert. *C. occidens* from the deserts of southwestern USA is a striking species in which the abdomen is bright orange, and the glossy black carapace has a white median band.
DISTRIBUTION More than 25 species in the USA, many of them very widespread.

ZUNIGA MAGNA

COMMON NAME Ant spider
DESCRIPTION The top of the legs, thorax, and abdomen of this glossy black ½ in-/ 13 mm-long spider are partially clothed in a pelt of short golden hairs. Down the center of the abdomen there is a line of five white spots. The top of the head is very flattened, and the jaws project forwards. In common with most spiders that mimic ants, the front legs are held up and waved about as the spider walks around, thereby mimicking the constantly quivering antennae of an ant.
DISTRIBUTION Rainforests of eastern Brazil.

ANYPHAENA ACCENTUATA

COMMON NAME Buzzing spider
DESCRIPTION Both the female and the male (illustrated) are straw colored. In the male the sides of the abdomen and carapace are dark with four black spots in the middle of the rear half of the abdomen. The female is fatter and appears much paler, as her abdomen lacks the dark sides. This species is an active hunter on the foliage of trees, especially oaks. During courtship the male produces a buzzing sound by rapidly tapping the tip of his abdomen against a leaf. The females lay their eggs inside a nest consisting of several leaves fastened together with silk. Around 20 species of *Anyphaena* are found in the USA, most of which live under stones or logs, or among grasses and bushes.

DISTRIBUTION Occurs over Europe and Asia.

Stone Spiders *Family Gnaphosidae*

More than 2,000 species of mainly brown or black stone spiders are found worldwide.
The front spinners are rather long and tubular, and are widely separated. The eight eyes are
very small, befitting spiders that are mainly nocturnal hunters relying on touch and smell,
rather than sight, to locate prey. Most species spend the day under logs or stones, hence
the common name, emerging at night to hunt and mate.

SCOTOPHAEUS BLACKWALLI

COMMON NAME Mouse spider
DESCRIPTION With its furry brown
appearance and darting movements this
³/₈ in-/9 mm-long spider rather resembles a
small mouse. The fur has a somewhat greasy
sheen and the dark brown carapace is
narrowed at the front end. The spinners
protrude from the rear of the abdomen as
two small knobs. The mouse spider is
usually seen wandering slowly around on
the walls of houses, searching for prey.
The female in the illustration is guarding
her white egg-sac under a stone.
DISTRIBUTION Europe and most of Asia,
and in North America.

DRASSODES LAPIDOSUS

COMMON NAME Stone spider
DESCRIPTION The carapace of the ⁹/₁₆ in-/14 mm-long female (illustrated) is a grayish- to reddish-brown, with a black marginal line, and is densely carpeted with short hairs. The plump, egg-shaped abdomen is similar but often has a dark band down the middle, (this band is absent in the spider illustrated). The jaws are shiny, brownish black, and rather large. The male is similar to the female, but ⅛ in/3 mm shorter. The spider spends the day in a silken sac or tube beneath a stone or loose bark, emerging at night to hunt insects. There are six species of *Drassodes* in the USA. The common *D. neglectus* is a yellow or light gray spider with indistinct chevrons on the rear half of the abdomen.
DISTRIBUTION Europe, eastward to Japan, and in North Africa.

ZELOTES APRICORUM

COMMON NAME Black zipper
DESCRIPTION This glossy black spider is one of many similar-looking ¼ in-/6 mm- to ³/₈ in-/9 mm-long species found in Europe (more than 50 species in France alone), and the USA (around 30 species, many of them dark brown rather than black). The carapace is noticeably narrowed towards the front, so that the eyes are crowded into rather short rows. These spiders live under logs and stones, emerging at night to hunt.
DISTRIBUTION Europe and most of temperate Asia.

Huntsman Spiders *Family Sparassidae*
This is a mainly tropical family of generally large, flattened spiders with long legs. They are often called giant crab spiders, because their legs are held out crablike at their sides. Unlike true crab spiders (Thomisidae) they have teeth on their jaws. Sometimes the term Heteropodidae is used for this family.

MICROMMATA VIRESCENS

COMMON NAME Green meadow spider
DESCRIPTION The female (illustrated) reaches a length of ⁹/₁₆ in/14 mm and often sits and waits on leaves to ambush prey. The legs and carapace are a deep bright green, but the abdomen is yellowish green, and there is a deeper green cardiac band near its base. The male only measures ³/₈ in/9 mm and is a very beautiful spider, quite different from the female. His carapace and legs are a rather somber dull green but his slim abdomen is gold on top with a broad, red central band and red sides.
DISTRIBUTION Europe, eastward through Asia to Japan.

HOLCONIA IMMANIS

COMMON NAME Giant huntsman
DESCRIPTION This large spider is common in houses but it also lives on trees in forests. The overall color is grayish brown, with the legs banded dark and light, and a deep brown cardiac stripe extending from the front of the abdomen to about halfway down its length. It is a powerful hunter, and will tackle large insects, other spiders, and even lizards. Prey up to 4 in/10 cm long is taken by the 1³/₈ in-/34 mm-long females. The male is smaller (1 in/25 mm) and lives with the female for a while during the mating season. Humans are occasionally bitten but the bite is not dangerous, although rather painful.
DISTRIBUTION Native to Australia but introduced to New Zealand.

PANDERCETES GRACILIS

COMMON NAME Lichen huntsman
DESCRIPTION The flattened body is mottled gray and brown, while the legs are adorned with flattened tufts of hairs. The coloration helps the spider to blend into the lichen-speckled tree bark on which it spends the day, while the hairiness of the body and legs helps to eliminate shadows. The female (illustrated) reaches a length of ³/₄ in/19 mm, and guards her flattened disk-shaped egg-sac. She makes an attempt at camouflaging this by incorporating a few flakes of lichen into the tough silk on its surface.
DISTRIBUTION Australia (tropical Queensland) and New Guinea.

Wall Crab Spiders *Family Selenopidae*
This is a small family containing some 400 species of very flattened, gray or brown spiders
which live under stones or on rocks. The arrangement of the eyes, six in a single row,
is characteristic of the family.

SELENOPS RADIATUS

COMMON NAME Wall crab spider
DESCRIPTION This is one of only four species
of the genus found in the USA. The spider
illustrated was found in the Arizona desert.
The body and legs are heavily mottled in
dark and light brown, making the spider
very difficult to see when it flattens itself
against some desert rock-face. While waiting
for a meal to turn up the spider sits head-
downward but will sprint toward a crevice
with remarkable speed if disturbed. Several
species are common in houses, living behind
picture frames, fridges, furniture or toilets.
DISTRIBUTION Widespread in warm regions,
including southern USA.

Two-Tailed Spiders *Family Hersiliidae*
This family contains some 75 mainly tropical spiders which live on tree trunks and rock-faces. They are easily recognized by the rather triangular abdomen, and the pair of long spinners which protrude from the rear end, like two tails.

HERSILIA BICORNUTUS

COMMON NAME Two-tailed spider
DESCRIPTION As with all two-tailed spiders, the brownish or grayish mottled coloration makes the spider very difficult to spot when it is sitting head-downward on a tree trunk. The lens-shaped egg-sacs are covered in white silk, which the female camouflages with bits of bark or lichen prized off with her chelicerae. When an insect passes by, the spider fastens it to the trunk by running in circles around it, trailing silk from the long spinners. These spiders can often be found in the less dense parts of forests, such as along highways or around the edges of clearings.
DISTRIBUTION Southern Europe and Africa.

Crab Spiders *Families Philodromidae and Thomisidae*
The spiders in these two families are crablike in both shape and movement, walking forward, sideways or backward with equal facility. The males are much smaller than the females but have much longer legs in proportion to their bodies. Crab spiders do not build prey-catching webs but instead sit in ambush on flowers, leaves, bark, rocks or soil. More than 2,000 species have been described, mainly from the tropics.

PHILODROMUS DISPAR

COMMON NAME House crab spider
DESCRIPTION This philodromid can often be seen wandering around inside houses. It is also common on trees in woodland and on shrubs in gardens. The male (illustrated) is a slim-bodied spider with an iridescent black carapace and abdomen, and white sides. The female looks like a completely different species. She has a plump brown abdomen, and a brown carapace with dark bands on either side, bordered by white margins. She can often be found with her legs spread out protectively across her white silken egg-sac tucked into the corner of a wall.
DISTRIBUTION Europe, through most of Asia, and in North America.

TIBELLUS OBLONGUS

COMMON NAME Grass spider
DESCRIPTION This long, thin philodromid often sits head-downward on grasses, where it is very hard to spot. The body and legs are straw colored, and there is a light brown band down the middle of the body from front to rear. A short way in from the tip of the abdomen there is a pair of black spots.

The females measure ⅜ in/9 mm, the males slightly less. The females can often be found sitting astride their silk-covered egg-sacs affixed to a plant stem.
DISTRIBUTION Europe, eastward to Japan, and in most of North America.

XYSTICUS GULOSUS

COMMON NAME Plain crab spider
DESCRIPTION This is one of more than
70 species found in the USA, with another
17 in Europe. Nearly all of these are
rather similar in size and coloration to the
⁵/₁₆ in-/8 mm-long female in the illustration.
In many species the abdomen is more
heavily marked with dark chevrons, while
one or two are spotted. In some species the
tiny, dark brown males spin a bridal veil
over the female's head and legs before
mating. The females stand guard over their
white egg-sacs attached to plants. Although
often found on flowers, most species of this
genus spend much of their time waiting in
ambush on leaves, bark, stones or sand.
DISTRIBUTION Most of the USA, excluding
the deserts of the southwest.

THOMISUS ONUSTUS

COMMON NAME Heather spider
DESCRIPTION The females of this handsome spider can be pink, pale yellow or white. The pink form is almost always found on pink flowers, especially heathers, but the white and yellow forms do not always choose a matching background. The female (illustrated) has a very plump and rather triangular abdomen with angular tubercles on the top corners. On the face there are two horn-like projections on either side of the eyes. The females reach a length of ⁵/₁₆ in/8 mm but the brownish-orange males are tiny.
DISTRIBUTION Europe, eastward to Japan, and in North Africa.

THANATUS FORMICINUS

COMMON NAME Diamond spider
DESCRIPTION The ground color of this long-legged philodromid is light brown. At the front of the abdomen there is a rather elongate, dark brown, diamond-shaped mark, edged with white. Down the center of the carapace there is a broad, pale brown band. The male measures ¼ in/6 mm in length, the female ⁷/₁₆ in/11 mm. The diamond spider hunts for prey on tree trunks, cliff-faces, and among grasses, and shrubs.
DISTRIBUTION Europe, eastward through Asia, and throughout most of North America.

MISUMENA VATIA

COMMON NAME Common flower spider
DESCRIPTION In the USA this spider is often known as the goldenrod spider from its habit of sitting on the yellow flowers of that name. The same individual is just as likely to be found on white flowers, because this spider can change from yellow to white and back again, according to the color of the flowers it chooses as a background. The sides of the abdomen are often striped with red. The small, dark brown male cannot change color. The venom is very potent and insect victims rapidly succumb to its effects.
DISTRIBUTION Europe, eastward to Japan, and in North Africa and North America.

PHRYNARACHNE RUGOSA

COMMON NAME Warty bird-dropping spider
DESCRIPTION The wrinkles and warts which cover this strange spider, allied to its shiny appearance and blotchy coloration, all conspire to make it a perfect mimic of a wet, recently fallen bird-dropping. The ³/₈ in-/ 9 mm-long female (illustrated) sits motionless for day after day in full view on the same leaf. She does not merely wait passively for prey to chance by, but actually emits a special manure-like scent that attracts certain types of flies to come within reach of her grasping front legs.
DISTRIBUTION Tropical Africa and Madagascar.

SYNEMA GLOBOSUM

COMMON NAME Gold leaf crab spider
DESCRIPTION The small, shiny, and rather rounded abdomen of the female can be black and red (as illustrated), black and gold, or black and yellow. The black pattern on the abdomen resembles a human face. The carapace is black, which contrasts with the pale brown eyes. The females reach a length of ¼ in/6 mm and are usually found on flowers. The small black males have a white bar on the abdomen. There are three species of *Synema* in the USA. In *S. parvula* the legs and carapace are yellowish orange and the abdomen yellow with a black tip. It is common in the eastern states.
DISTRIBUTION From southern Europe through temperate Asia to Japan, and in North Africa.

CAMARICUS FORMOSUS

COMMON NAME Hallowe'en crab spider
DESCRIPTION The rather domed carapace is
deep red, contrasting with the milky white
and rather ghostly eyes. The glossy abdomen
is longer than broad, with a black Hallowe'en-
type mask set against a white background,
and a red rear end. The legs are a translucent
grayish white, with a few black markings.
The female (illustrated) reaches a length of
³⁄₈ in/9 mm. This is one of a number of small,
glossy crab spiders that prey solely on ants.
DISTRIBUTION India to Java, and in Vietnam.

STEPHANOPIS ALTIFRONS

COMMON NAME Knobbly crab spider
DESCRIPTION The rough exterior of this
rather square-bodied crab spider perfectly
matches the texture of the bark on which it
normally lives. It often wedges itself into a
crack in the bark, making itself even less
conspicuous. The general coloration is
grayish brown. At night the spider comes
alive and wanders over the bark in search
of prey. The female, which reaches a length
of ⁷⁄₁₆ in/11 mm, hides her egg-sac inside a
crevice in the bark. Like the lichen huntsman
and two-tailed spiders, she camouflages it
with flakes of bark.
DISTRIBUTION Eastern Australia.

Jumping Spiders *Family Salticidae*

This is a huge family of more than 5,000 species, most of which are tropical. Many of the males sparkle with iridescent, jewel-like colors, and often look quite different from the much drabber females. Most species are rather small, the largest being only ⁷/₁₀ in/35 mm long. They have eight eyes. Two large ones face forward and can be focused very accurately from as far as 20 in/50 cm away. The other eyes are smaller, and help to detect movement and fix the prey's position. These spiders can leap up to 30 times their own length.

SALTICUS SCENICUS

COMMON NAME Zebra spider
DESCRIPTION This ¹/₄ in-/6 mm-long, black and white spider is usually seen stalking across the walls and windowsills of houses. It is most common around houses and backyards but is also at home on rocks and trees away from habitations. The female (illustrated) stands guard over her egg-sac, which is normally placed under a rock, plank, flowerpot or other object. The carapace is black with two white spots, while the black abdomen has a white frontal margin and two pairs of white, backward-pointing chevrons. The male's jaws are large and project well forward of the face.
DISTRIBUTION Europe, through most of Asia, and over most of North America.

PLEXIPPUS PAYKULLI

COMMON NAME Pantropical jumper
DESCRIPTION The very broad, square face
of this attractive little $^7/_{16}$ in-/11 mm-long
spider is marked with horizontal black and
white stripes. The carapace is mainly black,
with a pale central stripe, and the abdomen
is black with one or two central stripes,
sometimes partly broken up into dots and
with white edges. Jumping spiders will
often tackle prey far larger and fiercer than
themselves, as depicted in the illustration,
which shows a spider feeding on a katydid.
DISTRIBUTION In most of the warmer parts
of the world, including southern USA
(Georgia and Florida westward to Texas).

ERIS AURANTIUS

COMMON NAME Variable jumper
DESCRIPTION This species is very variable.
The body is heavily covered in iridescent

scales, so that the colors appear different
depending on the angle. The overall color
is an iridescent, blackish bronze. Each side
of the carapace is marked with a whitish-
orange band. The entire margin of the
abdomen bears a broad, orange band,
although this can sometimes be very pale,
as in the individual illustrated. On top of
the abdomen there are two or three pairs
of white spots, and usually two pairs of
orange spots, which are less conspicuous
in the male. Females reach a length of
$^7/_{16}$ in/11 mm; males are marginally smaller.
DISTRIBUTION Central America and southern
USA, from Florida to Arizona and
northward to Delaware and Illinois.

LYSSOMANES VIRIDIS

COMMON NAME Leaf jumper
DESCRIPTION The ⁷⁄₁₆ in-/11 mm-long female (illustrated) is broadly similar to the male. The legs and carapace are vivid green, with the eyes set within a rectangular patch containing stripes of red, brown, and white scales (sometimes just light brown). The slender, tapering abdomen is of a more brownish shade of green. The males have huge curved chelicerae which jut out well forwards of the face and are overtopped by the even longer stalks of the palps.
DISTRIBUTION Central America and the Caribbean, and southeastern USA (North Carolina southward to Florida and westward to Texas).

PHIDIPPUS JOHNSONI

COMMON NAME Johnson's jumper
DESCRIPTION The carapace is black in both
sexes of this attractive spider. In the males
the abdomen is a strikingly bright red, with
a black hind margin. In the ⅜ in-/9 mm-
long female (illustrated) the red is more
subdued, and only occurs as a band down
the center of the black abdomen, or along
its margins. The face has a small mustache
of white hairs. The legs are black, and in
the males they are adorned with substantial
tufts of hair.
DISTRIBUTION Over much of southern USA,
from North Dakota southward to Texas and
then across to the Pacific coast states.

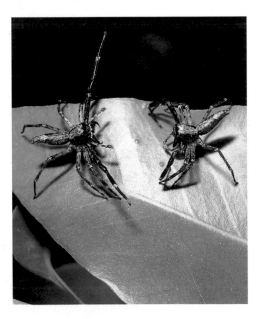

HELPIS MINITABUNDA

COMMON NAME Bronze Aussie jumper
DESCRIPTION In both sexes the abdomen is
long and slender, and the carapace broad
and high. The males are an iridescent,
bronze-brown, with very long black legs
with white joints. These long legs are used
in competitions between males, as depicted
in the illustration. The female is paler and
has shorter front legs. She is often found in
a nest, consisting of a silk sheet forming a
roof over a leaf.
DISTRIBUTION Eastern Australia and
New Zealand.

TELAMONIA DIMIDIATA

COMMON NAME Two-striped gaudy jumper
DESCRIPTION The male (illustrated) reaches a length of nearly ½ in/13 mm. His black front legs are adorned with dense fringes of black hairs. There is a red band across the eyes, and a white spot on top of the black carapace, which has broad white sides. The slender, tapering abdomen bears broad,

lengthwise bands of blackish-red and white. The slightly longer female looks completely different, with a pale whitish-orange ground color and two narrow orange stripes along the top of the abdomen. The carapace bears a number of orange blotches.
DISTRIBUTION India to Indonesia.

SNAKES

**Over 100 snakes identified,
with information on evolution,
movement, feeding, reproduction,
and distribution.**

INTRODUCTION

We think snakes are the most splendid of animals, though their alien form, so different from that of human beings, is loathsome to many people. They are successful predators who have pioneered elegant strategies to turn leglessness to their advantage. Some of them can catch a bat in mid air, even in the pitch dark. Others can climb among the thinnest of twigs, stretching and contorting their bodies into superb, geometric designs. There are some that are so strong that it takes three men or more to unwind their coils from around the trunk of a tree.

Being "snake fans" we also marvel at the diversity of their exquisite patterns and colors. There are over 2,700 species ranging in size from ½ in to 33 ft/10 mm to 10 m. Here we can only show you some of our favorites, but we hope this selection may inspire others to become interested in snakes.

WHAT IS A SNAKE?
Snakes are members of the class Reptilia together with lizards, crocodilians and turtles. All these reptiles have waterproof skins covered in scales or horny plates that can be modified to make elaborate crests or spines, or transformed into toughened armor or protective shells.

The White-lipped Pit Viper belongs to arguably the most advanced group of snakes.

Snakes show the most conservative of reptilian designs. Their streamlined bodies are stripped to the bare essentials and only a few have any adornments at all; some vipers have a horn on their snout and there is a water snake with tentacles, but as snake accoutrements go, that is about it. To identify one from another it is necessary to look at the marvellous patterns and colors of the skin, as well as subtle differences in shape and size. There are some species of snake so similar, that to tell them apart one must count the exact number of scales at specific points on the body.

Outside appearances may vary; a turtle is very different from a snake, for example, but on the inside all reptiles have an important feature in common. Unlike mammals and birds they do not have an internal heating system that allows them to maintain a constant body temperature. To keep active reptiles must rely on external sources of warmth, so if air temperatures are not high enough to keep them up to speed, they must bask in the sun. Under these conditions if you can watch a snake or lizard for long enough you will see it shuttling between sun and shade to keep its body at just the right temperature.

In the tropics it is so warm that reptiles can function for much of the time, but in temperate climates cool weather forces them to remain torpid in their retreats and they must hibernate during the winter.

Another feature that defines a reptile is that their embryos are enclosed within eggs, although some snakes and lizards may seem to give birth. Those able to do this retain their eggs within their bodies—becoming a sort of mobile incubator, until development is complete and they produce a litter of babies.

HOW SNAKES AROSE
When it comes to legless reptiles the snakes are not an exclusive club. Many lizards, some skinks, slowworms and the Australian Flap Footed lizards have lost or nearly lost their limbs. But why? These lizards are either burrrowers or denizens of confined spaces. In a tunnel anything that sticks out from the body can snag and hinder movement, so it is much easier to be a "smoothie" and do without legs. In all probability the first snakes evolved as legless creatures to be efficient burrowers.

Snakes have also lost their external ears and the ability to hear most airborne sounds. Instead the bones in their head are modified to detect vibrations, vitally important underground. A burrowing origin also led to snakes having their glassy, unblinking stare. They do not have eyelids, so can never shut their eyes, even when they are sleeping, but there is a tough transparent spectacle that protects the eye from damage when the snake is pushing through soil or dense vegetation. The prototype snake designed for a subterranean lifestyle probably arose from a lizard or lizard-like reptile. There are even some snakes that show direct evidence of an ancestor with limbs; the giant boas and pythons have tiny spurs at the base of the tail, the vestiges of legs. Many scientists think monitor lizards are the closest living relatives of snakes. Like them they have a deeply forked tongue for sampling odors in the air and the ability to swallow large items of food. The ancestral stock of snakes may have looked very similar to these giant lizards. Whatever may have happened 120 million years ago, the first snakes soon expanded their empire above ground and set about colonizing the world, using skills unique to themselves.

MOVING AROUND
At the surface snakes had to find effective ways of locomotion without the use of legs. They do this in three main ways:

1. "Squirming" or Serpentine movement
Muscular waves undulate the snake's body in a series of S shaped curves and it propels itself forward by levering the hind part of each curve against the irregularities in the ground. Snakes swim by pushing the curves against the resistance of the water.

2. "Caterpillar Crawl" or "Rectilinear Creeping"
A technique used by stout, heavy snakes such as pythons and vipers. They move slowly forward by pushing groups of belly scales against the ground while sliding others forward, giving the general impression of the whole body gliding in a straight line.

3. "Sidewinding"
Some snakes that live on shifting sand have a spectacular mode of travel. The snake

1

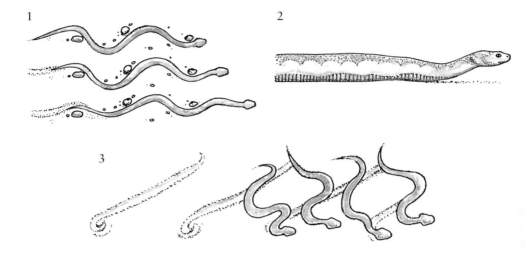

2

3

throws out a lateral arc with its head and the front part of its body; while transferring the rest of its bulk to this forward purchase, it throws out another loop, and by repeating this process it moves sideways in a series of steps.

SNAKE FEEDING

There are no vegetarian snakes; they are all carnivorous. It seems that for every type of animal food, whether frog spawn, spiders, slugs, rodents, birds, lizard eggs, even antelopes, there is a snake somewhere that is able to eat it.

These reptiles would seem to be at a disadvantage to other predators that can run and jump, but their thin bodies can be an asset. Have you ever seen a cat that could follow a mouse right into its burrow? Snakes have no claws for overpowering and tearing up prey but they have developed techniques that are uniquely their own. Some have a venomous bite and they all have the ability to swallow an item of food that exceeds the diameter of their heads. If it is a food that cannot fight back, it can simply be grabbed and held with the snake's inwardly curved teeth. These are useless for chewing, so the food must be swallowed whole. The bones in a snake jaw are joined by muscles and ligaments; this flexibility and elasticity allows a huge gape and ability to swallow larger items of food. The teeth are alternately freed and refastened and the food is dragged into the gullet. Then the neck muscles are brought into play, forcing the meal toward the stomach. The snake's skin is so flexible that it can be stretched without tearing if the food is really large. Really bulky prey can take a long time to swallow but the snake does not choke as it has a reinforced windpipe that can be thrust out of the side of the mouth, allowing it to breathe.

Not all prey can be swallowed alive—if it is large enough to cause injury or has

Boomslang: One of Africa's most venomous snakes.

teeth or claws, it must be immobilized first. To do this, snakes have their own specialized techniques. Boas, pythons, king snakes and several others use constriction. After an animal or bird has been grabbed, these snakes quickly throw one or two coils around it, drawing these inexorably tighter, until the prey loses consciousness because it is unable to breathe. The food is not crushed during this process, but is suffocated.

Other snakes can incapacitate prey with a single bite, as they have become living chemical factories specialized to produce toxic fluids. Venoms are modified digestive juices, clear or yellowish liquids whose exact composition varies between species. Some components cause the prey's nervous system to malfunction while others break down muscles and blood vessels. Most venoms cause a multitude of effects that work in concert to paralyse and eventually kill. The snakes store their venom in glands behind each eye that connect with teeth modified for injection. Vipers have the most elaborate fangs; when they are not in use they are folded in the roof of the mouth, but during a strike they swivel forward and venom is forced at high pressure through a duct in the tooth. Members of the cobra family have more simple fixed fangs that are generally shorter than those of vipers as they cannot be folded and stowed away.

Hijacking another creature with chemicals is one of the greatest snake accomplishments. This technique allows minimum physical contact with potentially dangerous prey. For instance it takes only a fraction of a second for a rattlesnake to strike, inject venom and release its hold. It will then wait or even retreat before tracking the scent trail of the dying animal and eating the body.

Of course, a weapon with the efficacy of venom can also be used in defense. But snakes will only waste this precious resource on larger creatures, including human beings, when they perceive their life is at risk. The rattle of rattlesnakes and the vivid color of coral snakes are warnings to keep away, so venom can be preserved for getting a meal. A bite in self-defense is always the last resort.

SNAKE COURTSHIP

Perfume is crucial in the courtship of snakes. During the breeding season the females produce a scent trail to draw in a suitor. A female in breeding condition exudes chemical messages or pheromones from her skin, so once a pair have made contact the male checks her out with a frenzy of tongue flicking. Courtship can be perfunctory, but a male can spend several hours using his chin to caress a female before he finally moves his tail beneath hers to mate.

For some species courtship is a highly gregarious affair, and a writhing "snake ball" is the result, with up to 30 males pushing and jostling around a female, until one of them succeeds in mating with her.

The males of some rattlesnakes, vipers and Australian elapids even indulge in

bouts of wrestling. These ritualized duels sometimes last for up to an hour as each of the combatants tries to gain the upper hand. The two snakes push and twist, raising their bodies above the ground, as each tries to force its opponent's head to the ground. They fall back repeatedly before trying again. Fatigue forces the weaker, usually smaller, snake to retreat and the victor mates with the female who is

Forest Cobra. Despite being a slender snake, this is Africa's largest cobra.

generally nearby, although she seems to take no interest in the combat dance, and could even mate with another passing male while the jousting is in progress.

SNAKE PARENTHOOD

After mating, female snakes produce between 3 and 16 eggs or young, although some have as many as 100. Development time depends upon species and temperatures but as a rule snake eggs take 2–4 months to hatch. Eggs are laid below the surface of the soil or sand and are hatched by the sun. A young snake escapes from the egg by slashing through the shell using an egg tooth on its snout. If it is warm embryos develop faster, so some snakes are more choosy in where they lay their clutch. Piles of manure or rotting vegetation generate heat when they decay and these are prime "hot spots" for the eggs of grass snakes and some rat snakes. Termite

Green Tree Python. Coils perfectly along a branch while waiting to ambush its prey.

mounds with their toughened walls of mud and insect saliva are both safe and warm, maintained at a constant high temperature because of the heat produced by the millions of termites; where these occur many snakes exploit them as a natural incubator. Few snakes guard their eggs, but there are exceptions. Many pythons become "broody," coiling around their clutch in a protective embrace until their young begin to hatch. To keep their eggs at a higher temperature than the surrounding air, some pythons even generate body heat by violently twitching their muscles.

The king cobra is the only snake that builds an elaborate nest. The female uses her snout to clear the nest site of stones or sharp twigs before using a coil to gather in leaves and grasses. Once she has collected enough nesting material she weaves in and out of the pile to form a more compact mound. She then lays her eggs in a roofed chamber on top, remaining on guard until the baby cobras hatch.

Other snakes are live bearers, retaining their eggs within their bodies, before giving birth to a litter of babies. Each youngster is individually wrapped in its own egg membrane, and must use a tiny egg tooth to cut an exit in this transparent capsule.

Mothercare is over once there are young snakes on the scene, and the snakelings disperse and must find food and survive on their own.

SNAKES AND MAN

Throughout history and all around the world people have been enthralled by snakes. In some cultures they are sacred

creatures, believed to bring fertility and to possess the secret of immortality. It is likely that snakes were attributed with these powers because of their phallic form and the fact that they seem to be reborn with brighter colors and a more velvety feel every time they shed their skins. Skin shedding or sloughing is, in fact, the method used by snakes to accommodate growth and to replace their outer covering when it becomes abraded and worn. This happens 4 to 6 times in a year.

In other human societies snakes are the ultimate evil, demons from the underworld that are dealers in death. This fearsome reputation arose from their unblinking gaze, their sinuous rapid movements and above all from the extraordinary properties of venom. Before death a victim's limbs

Snakes have been both revered and loathed by man throughout history.

could swell grotesquely and he or she could even sweat blood from the pores, effects that were real and not imagined. But there is nothing supernatural about snakes; they just have elegant solutions to life without limbs.

Left to themselves, snakes are incapable of malevolence towards human beings; they never intentionally attack a person unless they are threatened themselves, either by accident or design. Unseen snakes can be trodden upon accidentally or, if they are noticed in the open, provoked by people trying to kill or harass them. Many snake bites could be avoided by a few simple precautions. In "snake territory" you should not go barefoot and care should be taken when putting your hands near rocks, burrows or dense vegetation that could be a reptilian retreat. A snake seen away from cover should not be molested, but left to slip away to go about its business, a business that is of great benefit to man.

In Maharashtra state in India in the small village of Shirala the people know the value of snakes. Every year in July there is the Festival of Nagpanchmi. Dancers, brass bands and a procession of decorated carts wend their way through the narrow village streets. People jostle at every vantage point eager to pay their respects to a beautiful creature. On every cart there are earthenware pots containing Indian cobras. Gently and with great care, handlers take out the snakes and show them to the crowd. The villagers know that these cobras control rats and mice that would otherwise consume their crops and spread disease. The snakes riding the carts will be released back into the fields where they were captured. Some have been caught, taken part in the procession and then been released unharmed every year for a decade. Cobras will only bite a person in self-defense and the people of Shirala regard them with caution and respect, not fear and hatred. Shouldn't all snakes be treated the same?

HOW TO USE THE IDENTIFIER

The species in this book are arranged in groups; blind snakes and thread snakes; pythons and boas; typical snakes; front-fanged snakes; vipers.

KEY TO SYMBOLS

Interesting facts about each snake have been provided by means of at-a-glance symbols, for which there are four categories. They denote whether a snake is:

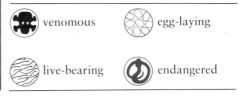

venomous   egg-laying

live-bearing   endangered

Trans-pecos Rat Snake. Primarily a nocturnal snake, feeding on birds, bats and lizards.

BLIND SNAKES (Family *Typhlopidae*)
THREAD SNAKES (Family *Leptotyphlopidae*)

Considered to be primitive snakes, there are about 175 species of blind snakes and 80 species of thread snakes in the warmer parts of the world. Most of them are small, although some attain a length of 2 ft/60 cm. All are burrowers and look more like worms than snakes, with small overlapping scales, tiny heads and minuscule eyes.

BLIND SNAKE
Leptotyphlops humilis

Otherwise known as a worm snake due to its obvious resemblance to the common earthworm.

DESCRIPTION Up to 16 in/41 cm, but usually smaller. The vestigial eyes appear as dark spots under the scales of the blunt head. A shiny snake, the color is purplish, brown or pink with a creamy underside.

DISTRIBUTION USA, southern California through to western Texas and into northwestern Mexico.

HABITAT Mostly found in arid areas but where there is some moisture in soil that is loose enough to burrow in. Occasionally found on the surface at night, but mostly dug out from under rocks, the roots of shrubs and around ant nests.

FOOD Feeds on small insects, especially ants, plus spiders, millipedes and centipedes.

BREEDING Lays a clutch of 2–6 tiny eggs.

FLOWERPOT SNAKE
Rhamphotyphlops braminus

A tiny snake, named after its habit of hiding in the soil in flowerpots and the like.

DESCRIPTION A large specimen would be 6 in/15 cm in length. These animals are black or chocolate with shiny scales, as if they have been polished. Tail ends in a spine.

DISTRIBUTION South East Asia, New Guinea and Northern Australia. Expanded its range by being accidentally introduced by man in containers of soil.

HABITAT Found in soil or under stones or logs, sometimes emerging at night through cracks in the floors of houses.

FOOD Worms, insects and their larvae.

BREEDING The only snake in the world to reproduce by virgin birth or parthenogenesis. All individuals are female and the eggs develop without fertilization.

PYTHONS (Family *Pythonidae* and BOAS (Family *Boidae*)

Includes all of the giant snakes. There are 37 species of boa and 27 species of python. They have vestiges of hind limbs that are seen as spurs at the base of the tail. They are primarily a tropical group although some small boas are found in more temperate climates. All subdue their prey by constriction. Pythons lay eggs; boas are live bearers. Many boas have a line of heat-sensitive pits along their lips so they can sense the body heat generated by their prey, even in pitch darkness.

BOA CONSTRICTOR
Boa constrictor

The classic constrictor often portrayed as a predator of man, yet these snakes are mainly placid and never grow big enough to consume a person.

DESCRIPTION To about 20 ft/6 m. A gray or silver snake with brown or deep red saddles along the back, though patterns and color tend to be geographically variable.

DISTRIBUTION Central Mexico, Central and South America down to Argentina.

HABITAT From semi-arid regions to, more commonly, rain forest habitat. Mainly lives in the tree tops, only coming to the ground to forage, occasionally attracted to human habitations by the availability of rodents.

FOOD A constrictor, can manage quite large mammals and birds.

BREEDING Live-bearing; 20–50 young measuring 12 in/30 cm in a litter.

DUMMERIL'S BOA
Acrantophis dumerili

Like most of Madagascar's wildlife, this snake is highly endangered.

DESCRIPTION Up to about 6 ft/2 m. Ground color a brownish-gray, with a richer red-brown on the back. The markings are deep brown, a thin line across the back widening into a fairly regular double-lobed mark with a white center to the lower part.

DISTRIBUTION Madagascar and the Mascarene Islands.

HABITAT Humid rain forests, sheltering in leaf litter, logs and mammal burrows. Hibernates during the dry, cool months from May to July

and breeds immediately upon emergence.
FOOD A constrictor, feeding on small mammals and birds.
BREEDING Live-bearing; 4–6 large young are produced.

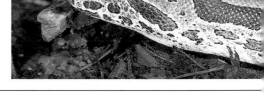

ANACONDA
Eunectes murinus

The heaviest snake in the world, can grow to a massive 286 lbs/130 kgs. The source of many myths, lengths of in excess of 60 ft/18 m have been claimed but a prize offered by the New York Zoological Society for a 30 ft/9 m specimen remains unclaimed.
DESCRIPTION It is speculated that this species can grow to more than 30 ft/9 m. Very heavy-bodied brownish-green to grayish-green snake, with ovoid markings in brown or black.
DISTRIBUTION Much of northern South America and Trinidad.

HABITAT Semi-aquatic; restricted to swamps, marshes and river valleys.
FOOD A constrictor capable of subduing and consuming prey the size of young tapirs. Often uses water to conceal itself from prey that includes mammals, aquatic birds, caymans and crocodiles.
BREEDING Live-bearing; between 10 and 50 young normally.

EMERALD TREE BOA
Corallus caninus

This snake gives an excellent example of convergent evolution as it is strikingly similar to the Australian green tree python, a snake that evolved in isolation thousands of miles away from this boa.

DESCRIPTION Bright green with a white or yellow belly and the back intermittently broken by bands and flecks of white. The young are similarly patterned, but a striking reddish-orange or bluish-green.

DISTRIBUTION The Amazon basin, Peru and Ecuador through Brazil and Bolivia to the Guianas.

HABITAT Totally arboreal, this snake uses its prehensile tail to grip branches while hanging loops of its body over the branches to be perfectly camouflaged during the day. Active at night, the adults forage high in the trees while the juveniles inhabit lower trees and bushes.

FOOD A constrictor, feeding on birds and mammals ambushed in the trees.

BREEDING Live-bearing, the young quickly making their way to lower parts of the trees.

Rainbow Boa
Epicrates cenchria cenchria

Perhaps the most beautiful snake in the world, not just because of its markings but mainly due to the remarkable iridescence of the skin.

DESCRIPTION A medium-sized boa growing to a maximum of just over 6 ft/2 m. The skin is reddish brown with darker lateral rings and spots, but with microscopic ridges on the scales that act like a prism to refract light.

DISTRIBUTION Through much of northern South America and into the Amazon Basin. Other rainbow boa sub-species range over most of South America.

HABITAT Found in forests, woodland and plains.

FOOD A constrictor, feeding on small mammals and birds.

BREEDING Live-bearing, producing litters of around 20 young.

PACIFIC ISLAND BOA
Candoia aspera

There are three species of Pacific Island boas; this is the smallest and plumpest.

DESCRIPTION Up to 3 ft/1 m in length. Triangular head with a characteristic straight-edged snout. Its keeled scales have a row of diamond shaped markings.

DISTRIBUTION New Guinea, Moluccas, Bismarck Archipelego, Solomon and Tokelau Islands.

HABITAT Spends time in the trees and is fond of bathing.

FOOD Lives on a diet of small mammals, birds and lizards.

BREEDING Gives birth to a litter of living young.

ROSY BOA
Lichanura trivirgata

A normally tame snake but its nocturnal habits were alarming enough to the settlers of the western states of North America that they inevitably led to many snakes suffering a terminal fate at the hands of a "six-shooter".
DESCRIPTION A heavy-bodied snake up to 44 in/110 cm. Beige or rosy above, with blotched creamy underparts, prominent anal spurs and with a small elliptically pupilled iris.
DISTRIBUTION Southern parts of California and Arizona, Baja California and northwestern Mexico.
HABITAT Mostly nocturnal, living in arid areas, but sometimes found near water sources. When molested coils into a ball and protects the head with its body coils.
FOOD A constrictor, eating small mammals and birds.

BREEDING Live-bearing; clutches of between 3 and 12 young born in October or November.

RUBBER BOA
Charina bottae

A disturbed boa may curl into such a tight, spherical ball that it will roll with surprising ease.
DESCRIPTION 14–33 in/35–83 cm long. A stout-bodied snake, with an extremely short and blunt tail. Its skin has a rubbery feel. The top of the head has large symmetrical scales; pupils vertically oval. It is plain brown to olive green above, yellow to cream below, usually with no pattern.
DISTRIBUTION Western North America. Mid

California to southern British Colombia, Pacific coast to central Wyoming. Wide altitude range, from sea-level up to 1,000 ft/3,000 m.
HABITAT A burrower found in grassland and woodland, especially below rotting logs and rocks.
FOOD Constrictor, eating young mice, shrews, salamanders and snakes.
BREEDING Live-bearing; 2–8 young born in later summer or early fall.

KENYAN SAND BOA
Eryx colubrinus loveridgii

An inoffensive snake, yet one that can suddenly and violently ambush prey, bursting out from beneath the sand to overpower the prey animal.
DESCRIPTION Up to about 38 in/95 cm. Usually a dark sandy-brown, with irregular blotches of dark brown on the back. occasionally black and silver individuals are found. The tail is very short and the head is small, with little in the way of a neck obvious
DISTRIBUTION Ranges through much of east Africa.

HABITAT Found in dry bush and semi desert. A strong burrower, rarely seen at the surface except when dug up in such as maize fields (inevitably these individuals are accidentally killed during the process).
FOOD A constrictor, eating lizards and small rodents.
BREEDING Live-bearing, giving birth to around 20 young.

CALABAR GROUND PYTHON
Calabaria reinhardtii

Local people know this creature as the snake with two heads because when it moves above ground it waves its tail in the air, presumably to distract predators from attacking its head.
DESCRIPTION Up to 3 ft/1 m in length. Its small head is not distinct from its cylindrical body. Brown to reddish brown, with reddish spots.
DISTRIBUTION West and central Africa.
HABITAT Like the sand boas this python is designed for burrowing, with a blunt head and smooth scales. It remains hidden in the day, amongst roots or in rotting logs, emerging at dusk to hunt. Curls into a ball when attacked. Inhabits rain forests and savannah.
FOOD Lizards and rodents.
BREEDING Little known. This snake is an egg layer, the number in each clutch being small.

BURMESE PYTHON
Python molorus

Snake dancers prefer this python above all others for their stage shows.

DESCRIPTION Reaches 26 ft/8 m in length. Markings are an intricate pattern of irregular brown blotches on a yellow background.

DISTRIBUTION Southern Asia. Pakistan in the west to southern China in the east, and south to the Malay Archipelago.

HABITAT Often found near water, this sluggish python waits patiently for prey either coiled on the ground or hanging below a branch. In the cooler parts of its range it becomes torpid during the winter. Bharatpur Reserve in India is famous for its "hibernating" pythons.

FOOD Subdues mammals and birds by constriction. There is even a record of a 20 ft/ 6 m python killing and eating a leopard.

BREEDING Lays as many as 100 eggs. The female coils around them until they hatch, twitching her muscles to generate heat if the air temperature becomes too cool.

ROYAL PYTHON
Python regius

Otherwise known as the "ball" python from its habit of rolling up into a tight ball when alarmed.

DESCRIPTION Usually no more than 5 ft/1.5 m. Smallest of the African pythons, it is stocky with a short tail. The ground color is brown or bluish-brown, with a pattern of ovoid blotches of variable color that normally includes yellow.

DISTRIBUTION West Africa, Sierra Leone, Togo, Senegal and Gambia.

HABITAT Forests and woodland, remaining dormant during the dry season, often holed up in tree crevices. Collected extensively for the pet and skin trades, these pythons are now considered threatened.

FOOD A constrictor, these snakes feed on small rodents and birds.

BREEDING Small clutches of between 2 and 7 eggs which the female broods under her coils for the 3 month incubation period.

RETICULATED PYTHON
Python reticulatus

The longest snake ever recorded, with one specimen found in Celebes in 1912 measuring 32 ft 9½ in (almost exactly 10 metres).
DESCRIPTION Up to about 33 ft/10 m, but relatively slim compared to its length. The striking colors and patterns allow the snake to blend into its surroundings.
DISTRIBUTION South East Asia, the Malay peninsula, Borneo, Java, Sumatra, Timor, Ceram and the Philippines.
HABITAT Rain forest and woodland, but most usually associated with rivers, lakes and their environs. A good swimmer, a fact that is reputed to account for its occurrence on many isolated islands.
FOOD A constrictor, feeding mostly on mammals and birds but also large lizards, like monitors, and snakes. There have also been rare reports of these snakes preying upon human beings.
BREEDING Lays eggs. As in most snakes, the size of the clutch will depend almost entirely on the size of the female.

BLACK-HEADED PYTHON
Aspidites melanocephalus

In cool weather this python will stick out its head from shelter so that the black scales absorb the heat of the sun.

DESCRIPTION Attains 6 ft/2 m or more. Its head is jet black, while its body is yellowish or reddish brown with dark brown or black cross bands.

DISTRIBUTION Northern half of Australia.

HABITAT Found in humid coastal forests, dryish woodlands and grasslands. By day these pythons hide in burrows, hollow logs among tree roots, or in cavities in termite mounds. At night they hunt by probing into the retreats of their prey.

FOOD Ground-nesting birds, small mammals and reptiles including venomous snakes.

BREEDING Males indulge in ritualized combat during the breeding season. Female lays 5–12 eggs which she embraces with her body to guard until hatching.

BISMARCK RINGED PYTHON
Liasis boa

A relatively little-known snake, named for its geographical location rather than any relationship with the German statesman of the late nineteenth century.

DESCRIPTION Under 6 ft/2 m. The adults are poorly marked with black rings or are uniformly blackish-brown, usually with a light mark behind the eye. The juveniles are by far the more spectacular, being ringed in orange and black, a coloration that fades as the snake matures.

DISTRIBUTION The Bismarck Archipelago, Papua New Guinea.

HABITAT Found in rain forests, foraging actively for prey at night.

FOOD Constrictor, preying mainly on small rodents.

BREEDING Around a dozen eggs are brooded by the female.

CARPET PYTHON
Morelia spilotes variegata

Australia's best-known python, with a variety of distinctive subspecies. Can reach a length of 13 ft/4 m, although 8 ft/2.5 m is the average.

DESCRIPTION All of the forms have a dark ground color, blotched or patched with yellow or white markings which are bordered with black. The diamond python is the most distinctive and beautiful subspecies. The center of each of the carpet python's dorsal scales is bright yellow or cream.

DISTRIBUTION Most of Australia except for the arid center and the west.

HABITAT Equally at home on the ground or in the trees. Found in woodland, rocky outcrops and ravines, along water courses and in towns. Can even enter aviaries or chicken runs in search of food.

FOOD Small mammals and birds, although younger snakes prefer lizards.

BREEDING An average clutch contains 12–25 eggs. The female coils around them until they hatch, "shivering" her muscles to warm them if the temperature drops.

AMETHYSTINE PYTHON
Morelia amethistina

One of Australia's longest snakes, one specimen was measured and found to be 28 ft/8.5 m long.

DESCRIPTION Usually grows up to 23 ft/ 7 m. A slender python with a prominent "neck." Olive brown in color with bands of black that are usually broken up into irregular blotches.

DISTRIBUTION New Guinea and the northern part of Australia.

HABITAT Found in rain forests, savannah, woodland, even amongst scrubby vegetation on coral islets. At night these pythons either hang by their prehensile tails waiting to ambush prey or actively search for food in burrows, rock crevices or caves.

FOOD Rats, bandicoots, possums, fruit bats, wallabies and birds.

BREEDING The female lays 7–19 eggs which she coils herself around until they hatch about 3 months later.

GREEN TREE PYTHON
Chondropython viridis

Rests and waits for prey by folding its body along a branch in a series of perfectly measured loops, so that its head is always at the center.

DESCRIPTION This snake can be 6 ft/2 m long. Hatchlings are lemon yellow, getting the adult coloration at 1–3 years of age, which is an emerald green with the scales along the spine being white, yellow or cream.

DISTRIBUTION New Guinea and tip of north-eastern Australia.

HABITAT A denizen of humid rain forests, coiling in its characteristic fashion during the day and moving about at night.

FOOD Mammals, and especially birds, which it catches from ambush sites.

BREEDING 11–25 eggs that the female coils around, until hatching in about 50 days.

ROUGH SCALED PYTHON
Morelia carinata

Described by science in 1981, the first photographs of a living specimen were taken in 1992 by John Weigal of the Australian Reptile Park after an adventurous quest into unmapped territory.

DESCRIPTION Grows to 6 ft/2 m in length. It is brown in color and is unique among pythons in having keeled scales.

DISTRIBUTION Northern Kimberley District of Western Australia.

HABITAT Only a handful have been found in rainforest gorges, some amongst rocks, others in the trees.

FOOD Probably feeds on small mammals and birds.

BREEDING Unknown.

WATER PYTHON
Liasis fuscus

This snake can be so abundant there may be up to a ton of pythons in a square kilometre.

DESCRIPTION Usually 5 ft/1.5 m long, although some individuals are 10 ft/3 m. Blackish brown in color, with a shiny appearance and fantastic iridescent scales.

DISTRIBUTION Northern Australia, New Guinea and the Lesser Sunda Islands.

HABITAT Found near rivers, swamps and billabongs. In the warmer part of the year spends most of its time in the water.

FOOD Small mammals including wallabies and water birds. Where it is extremely abundant the huge population is sustained primarily by dusky rats.

BREEDING Lays about a dozen eggs which the female coils around until hatching, shivering to warm them, if temperatures drop too low.

TYPICAL SNAKES (Family *Colubridae*)

Typical snakes contains about three quarters of all types of snakes, over 1,500 species. More detailed taxonomic studies are beginning to subdivide this "rag bag" group; for example, many authorities split off file snakes into a separate family. Most Colubrids are totally harmless to people, although some are not. The Twig snake (*Thelotornis kirtlandii*) and Boomslang (*Dispholidus typus*) have fangs at the back of their jaws and a potent venom.

CORN SNAKE
Elaphe guttata

Possibly the most popular captive snake, with a remarkable number of strains bred to enhance the differing colorations.

DESCRIPTION 2–6 ft/60–180 cm, a slender snake with a wide variety of colors but mainly from orange to gray with strong brown to red markings. First blotch on neck divides into two branches that extend forward to produce a spearpoint ending between the eyes.

Usually exhibits a diagonal eyestripe and a checkerboard patterning against cream on the underside.

DISTRIBUTION Widely distributed throughout the southern and eastern USA and northeastern Mexico, with a few isolated populations outside that range.

HABITAT A strong climber, but mainly terrestrial. Occasionally found on farmland but more usually in woodland and on rocky slopes. Can be difficult to locate due to its propensity for hunting and resting underground in rodent burrows.

FOOD A constrictor, it eats small mammals, birds, lizards and frogs.

BREEDING Lays a clutch of 3–21 eggs from May to July.

GREEN RAT SNAKE
Elaphe triapsis

Elusive and poorly known but can be identified, even in poor light, by the fact that its head is longer than other American rat snakes.

DESCRIPTION 2–4 ft/60–125 cm. Slimly built, green or olive with no markings above and cream or whitish underparts. The scales are mostly weakly keeled, while the anal scale is divided.

DISTRIBUTION America, extreme south of Arizona and New Mexico through Mexico and into Guatemala and Costa Rica.

HABITAT Crepuscular and partially arboreal, it retires into rock crevices and underground at night.

FOOD A constrictor seeking out birds and their nestlings, but taking other prey, including small rodents and lizards.

BREEDING A clutch of in excess of 6 eggs is laid under stones or in similar locations.

BLACK RAT SNAKE
Elaphe obsoleta obsoleta

Otherwise known as the pilot snake, since it is believed to direct rattlesnakes and copperheads to hibernation sites.
DESCRIPTION A shiny black snake between 3½ ft/106 cm and an exceptional 8½ ft/256 cm. Body shape in cross section is rectangular rather than round.
DISTRIBUTION Central and eastern USA, as far north as New York and Ontario, down to Louisiana and Oklahoma.
HABITAT A strong climber that may take up residence in tree cavities. Large numbers may congregate at traditional hibernation sites, often in the company of rattlesnakes or copperheads.
FOOD A constrictor, feeding on birds, small rodents and eggs.
BREEDING Lays up to 36 eggs in decomposing plant material.

TEXAS RAT SNAKE
Elaphe obsoleta lindheimeri

Known as the "meanest" American rat snake, this snake is often so aggressive that it never calms down, even after many years of captivity.
DESCRIPTION 3½–7 ft/106–218 cm. The brownish or bluish-black blotches are not heavily contrasted with the gray or yellowish ground color. The head is often black, and red can be found on the skin between scales. There is a fair amount of variation in this species; however, the scales are keeled in all forms and the anal scale is divided.
DISTRIBUTION From the Mississippi Basin, west through Louisiana into central and southern Texas.
HABITAT USA, from swamps through to drier, rocky lands in the west of its range. Often found dead on the side of highways due to its habit of basking on the road surface.
FOOD A constrictor, feeding on small rodents, birds and eggs.
BREEDING The clutch of 6–28 eggs is laid from June to August.

Gray Rat Snake
Elaphe obsoleta spiloides

This snake is often associated with oak woodlands, from where it gets its alternative name, the "oak" snake.
DESCRIPTION 3½–7 ft/106–214 cm. This species is strongly blotched with, unlike most other rat snakes, similar markings to those of the juvenile. Blotches may be brown or gray, with a ground color varying between gray, pale brown or nearly white. Scales are weakly keeled and the anal scale is divided.
DISTRIBUTION USA, from Georgia to Mississippi and in a band northward to southern Indiana and Illinois.

HABITAT Found in farmland, on rocky slopes and in woodland. Where the species overlap it has a tendency to hybridize (or intergrade) with the black rat snake.
FOOD A constrictor, eating mainly small mammals, but it is also a favored food of many hawks.
BREEDING Clutches of less than 30 smooth-shelled eggs are common.

TRANS-PECOS RAT SNAKE
Elaphe subocularis

Easily recognized, since it is the most "bug-eyed" snake within its range.
DESCRIPTION 3–5½ ft/86–168 cm, a yellowish olive to tan snake, with diagnostic dark H markings along the back. The head is broad for a rat snake and the eyes are particularly large and protrusive with an extra layer of small scales beneath them.
DISTRIBUTION Southern New Mexico through western Texas down to northern Central Mexico, mostly associated with the Chihuahuan Desert.
HABITAT Primarily nocturnal in arid or semi-arid locations, especially in rocky areas at an altitude of between 1,500–5,000 ft/ 500–1,500 m.
FOOD A constrictor feeding on rodents, bats, birds and lizards.
BREEDING Lays a clutch of 3–7 soft, leathery eggs in summer.

LADDER SNAKE
Elaphe scalaris

This snake's name refers to the H shape markings that run along the back of juveniles but normally fade by the time the snake is adult.
DESCRIPTION Up to 5 ft/160 cm but normally less than 4 ft/120 cm. Large with smooth scales, an overhanging snout and a short tail. Adults are yellow-gray to brown, with a pair of dark brown stripes on the back; the belly is whitish or yellow but variably marked with black.
DISTRIBUTION Southwest Europe, most of Iberia, the Mediterranean coast of France and Minorca.
HABITAT Mostly diurnal, preferring stony habitats it can be often found in vineyards or around dry-stone walls. It can climb well and tends to be very aggressive if captured.
FOOD Constricts large prey, like a small rabbit, but eats variously sized mammals, nestling birds and grasshoppers, when young.
BREEDING Lays 6–12 eggs of about 2 in/5 cm in length, in and around July.

LADDER SNAKE

FOUR-LINED SNAKE
Elaphe quatuorlineata

This is the longest snake in Europe, but its name can give the wrong impression as only the hatchlings and juveniles exhibit the four-lined pattern.

DESCRIPTION Up to 8 ft/250 cm, but generally under 5 ft/150 cm. Large with a long, slightly pointed, head and keeled scales that lend it a rather rough appearance. The most robust snake in its geographic region, it can also be distinguished by the presence of two preocular scales (directly in front of the eyes). Color and pattern vary greatly, and the 4-lined markings fade as the snake gets older, but the belly is mainly an olivish yellow.

DISTRIBUTION Southeastern Europe into Russia, Italy, Sicily, many of the Aegean islands and South West Asia.

HABITAT Prefers humid areas near water, climbing and swimming well, often in overcast conditions or at dusk.

FOOD Constrictor, feeding on mammals, birds, eggs and lizards.

BREEDING May lay a clutch of around 20 eggs.

FOX SNAKE
Elaphe vulpina

The name fox snake originates from the fox-like odor that is associated with the fluid it discharges when threatened.

DESCRIPTION 3–6 ft/91–179 cm long. Heavily blotched on the body and with a ground color of yellowish to light brown; sometimes the head may be red-orange which can lead to its mis-identification as a copperhead.

DISTRIBUTION America, from southern Ontario south, between Indiana and Nebraska.

HABITAT Found in marshland, grass prairie, farmland and in riverine woodland. This snake is abundant in the marshy areas and dunes of the Great Lakes region and large numbers may congregate at hibernation sites.

FOOD A constrictor, feeds on rodents, frogs, birds and their eggs.

BREEDING 6–29 leathery eggs are deposited from late June to early August.

SINALOAN MILK SNAKE
Lampropeltis triangulum sinaloae

The name milk snake came from the belief that they sucked the milk from the udders of cows. One of the 25 subspecies of the milk snake, this snake was only described by scientists in 1978.
DESCRIPTION 3½–4 ft/102–122 cm. Head black,with some mottling of white, usually around the snout. The first black ring usually touches the angle of the jaw and creates a V shape on the throat; the red scales are not tipped black while the white scales are. There are between 10 and 16 red rings that are about three times the width of the black-white-black rings. All the body rings completely encircle the snake.
DISTRIBUTION Throughout Sinaloa, extending into neighbouring Mexican states.
HABITAT Little is known yet of this snake's natural history other than that it is found below 1,000 metres (3,000ft) and often around cornfields.

FOOD Small rodents, lizards and perhaps some invertebrates.
BREEDING Clutches are usually small, between 2 and 16.

HONDURAN MILK SNAKE
Lampropeltis triangulum hondurensis

It is thought by many that the striking "tri-color" markings of milk snakes evolved originally as a form of mimicry of the highly venomous coral snake.
DESCRIPTION Up to 4 ft/120 cm. The head is black but with a distinct yellowish band on the snout and a second band that broadens as it makes its way from the top of the head. It has red-orange scales that may or may not be tipped with black. The body rings of red, black and yellow entirely encircle the body: there are 13–26 red body rings. Some specimens may lack yellow rings and have a generally dark or tangerine appearance.

DISTRIBUTION Much of Honduras, Nicaragua and possibly northeastern Costa Rica.
HABITAT Found at lower elevations, usually located under rotting logs or stumps, it is most active at night.
FOOD Able to subdue and eat a variety of other snakes, it also feeds on small lizards, mammals and birds.
BREEDING As with other milk snakes the eggs are remarkably long and cylindrical; deposited in rotting substrate, they hatch out around August.

PUEBLAN MILK SNAKE
Lampropeltis triangulum campbelli

Alongside Dixon's milk snake, this is the most recently discovered milk snake (1983).
DESCRIPTION 2–3 ft/71–91 cm. A distinctive snake, with broad white body bands and a white mottled snout. The tail has around 5 black and white bands with no red-orange bands; the average body rings are 16 white and red bands and 32 black. About half the red bands are not complete on the underside, while there is no black tipping of the white scales.

DISTRIBUTION Restricted to a small area of southern Mexico.
HABITAT Prefers arid areas at reasonable elevation, from 5,000 ft/1,500 m.
FOOD Small rodents, snakes and lizards are the most common prey items.
BREEDING Rarely deposits more than 14 elongated eggs, which have an incubation period of 6–9 weeks.

SCARLET KING SNAKE
Lampropeltis triangulum elapsoides

If you can remember the old rhyme, "Red and yellow kill a fellow . . ." then you can be sure that the snake you are identifying is or is not the highly venomous coral snake but the harmless scarlet king snake.
DESCRIPTION Up to 2¼ ft/68 cm. A mimic of the eastern coral snake, but the tip of the snout is red and the yellow rings are separated from the red by black. The yellow bands may sometimes be white; all bands generally continue across the belly.
DISTRIBUTION North America, from southeastern Virginia through much of the eastern USA down to the tip of Florida, west to the Mississippi.
HABITAT Especially fond of pine woodland, hiding behind bark or underneath logs and often wintering in tree stumps.
FOOD A constrictor, feeds on lizards, snakes, baby rodents, fish, insects and earthworms.
BREEDING 2–15 eggs are laid in June or July.

GRAY-BANDED KING SNAKE
Lampropeltis alterna

Was once considered to be a rare snake, but is actually quite common, its nocturnal habits accounted for the fact that it was rarely seen.
DESCRIPTION Up to 5 ft/150 cm. Has quite a distinct head shape, gray with black lines or dots. Its pattern consists of a series of white-edged, red-centered, black blotches, the rest of the body being gray. There are between 9 and 39 black blotches or saddles, giving an indication of how variable a snake this can be. A distinctive characteristic is the relatively large eye with its silvery-gray iris.

DISTRIBUTION Southern Texas and into northern Mexico.
HABITAT Partial to arid to semi-humid habitats ranging from desert through to mountains. Nocturnal and secretive, once believed rare but now known to be abundant, it is able to pass an evil-smelling musk when captured.
FOOD Almost exclusively feeds on lizards, but will occasionally take small rodents.
BREEDING Small clutches of eggs are generally deposited beneath stones.

CALIFORNIA KING SNAKE
Lampropeltis getulus californiae

Like many of the king snakes, this snake is relatively immune to the venom of rattlesnakes.
DESCRIPTION The ground color for this snake is brown or black with stripes or rings of white or cream. The ringed form is the most common throughout its range.
DISTRIBUTION Ranges down the west coast of America from Oregon down through Baja California, and its eastward boundaries run from Nevada down into Mexico.

HABITAT A wide range of habitats, from rivers and grassland to desert or forest. Mostly diurnal, but more crepuscular in arid areas.
FOOD A constrictor, eating eggs, lizards, birds, amphibians, rodents and snakes, including rattlesnakes.
BREEDING 6–24 eggs in a clutch that may be deposited in rotting logs.

GREEN WATER SNAKE
Nerodia cyclopian

Easily mistaken for the highly venomous cottonmouth.

DESCRIPTION A heavy-bodied snake between 2½–4 ft/76–127 cm. Identification is difficult, few markings being distinguishable on the greenish or brownish back, but best recognized by a row of scales that are present between the eye and the lip scales.

DISTRIBUTION USA, primarily in the Mississippi Valley.

HABITAT A diurnal species, often found basking on low tree limbs near water. They can be found in great numbers in undisturbed areas, normally near quiet waters.

FOOD Primarily eating small fish such as minnows.

BREEDING Live-bearing; perhaps over 100 young produced in a litter.

SOUTHERN WATER SNAKE
Nerodia fasciata

Usually the most placid of the water snakes, often only regurgitating their last meal upon any assailant before making their escape.
DESCRIPTION 2–5 ft/61–152 cm. Can be identified by the dark stripe from the eye to the angle of the jaw, spots at the side of the belly and dark bands across the back. Colors usually darken with age, even to black, but can vary between gray, tan and red with red, brown or black bands.
DISTRIBUTION USA, in a coastal band from North Carolina to Alabama.
HABITAT Found in virtually every freshwater habitat from slow-moving streams through to marshes and even into saltwater regions.
FOOD Salamanders, frogs and small fish.
BREEDING Live-bearing; litters of up to 57 young are born from June to August.

GRASS SNAKE

GRASS SNAKE
Natrix natrix

There are some populations of black (melanistic) grass snakes and some individuals that play dead when attacked.

DESCRIPTION Normally up to 4 ft/120 cm but sometimes as much as 6½ ft/200 cm. Most specimens have a characteristic collar of yellow or white; the body is olive-gray, greenish to even silvery-gray with dark blotches and stripes.

DISTRIBUTION Virtually all of Europe below the Arctic Circle, across into Russia and south into Iran and Iraq.

HABITAT Marshes, meadowlands, farmland and hillside adjacent to rivers. Famous for its habit of voiding the contents of its anal gland when handled and can even feign death when threatened.

FOOD Feeds mainly on frogs and toads, but also takes fish, tadpoles, newts and even small mammals.

BREEDING Females can retain their eggs for up to 2 months; shortening the incubation period has allowed this snake to extend its range north. Deposits its eggs in decomposing plant material; often more than one female will use a single nest site.

DICE SNAKE
Natrix tessellata

An extremely aquatic snake, spending more of its time in water than any other European water snake.

DESCRIPTION Up to around 3½ ft/100 cm. It has a rather small, pointed head, but can also be distinguished by the pattern of dark square markings that lend the snake its name. The ground color can vary between grayish and brownish-green; some populations may be black or even yellow.

DISTRIBUTION Southeastern Europe to Afghanistan, Pakistan and even into China.

HABITAT Always found close to or in water. May climb small trees, but if disturbed will immediately drop into the water to escape.

FOOD Almost totally piscivorous but will take amphibians.

BREEDING Having mated after hibernation, the females lay up to 24 eggs under rotted logs or stones.

VIPERINE SNAKE
Natrix maura

This snake can be confused with a viper due to the presence of zigzag markings on the back, but if seen in water it will almost always be safe to assume it is the viperine snake.

DESCRIPTION Normally reaches around 2½ ft/ 70 cm in length. Usually exhibits some dark markings on the small, but broad, head. Individuals are usually brown or gray with a pattern of dark markings down the back and dark blotches on the side, mainly with light centers.

DISTRIBUTION Southwestern Europe, Sardinia and parts of North Africa.

HABITAT Diurnal, chiefly found in or around water, preferring weedy ponds and rivers. Dives readily when disturbed; if cornered will strike repeatedly, but with its mouth closed.

FOOD Eats frogs, toads, newts, tadpoles, fish and even earthworms.

BREEDING 5–20 eggs are laid in June to July, and hidden under rocks, in sand or in decaying vegetation.

COMMON GARTER SNAKE
Thamnophis sirtalis

The most widely distributed snake in North America.

DESCRIPTION 1½–4 ft/46–124 cm long. Characteristically has 3 yellowish lateral stripes and a double row of spots between the stripes that may actually predominate in some individuals. However, this snake is extremely variable; some specimens are virtually stripeless, black, green, brown and olive being among the ground colors individuals may exhibit. The scales are keeled and, like most garter snakes, it is distinguished from the water snakes by having a single anal scale.

DISTRIBUTION Southern Canada to the Gulf coast and west to California, only missing from the desert regions of southwest North America.

HABITAT A common snake found in woodland, marshes, along rivers and drainage ditches and even in city parks.

FOOD Mainly eats amphibians, tadpoles and earthworms.

BREEDING Live-bearing, producing 7–85 young between June and October.

RED-SIDED GARTER SNAKE
Thamnophis sirtalis parietalis

The famous inhabitant of Canada's snake dens.
In areas where hibernation sites are at a
premium, thousands of snakes gather at prime
sites.
DESCRIPTION Grows to about 2 ft/60 cm long.
Black in color with a strip of yellow along its
back and two yellow stripes on its side. It gets
its name from the red bars between the back
and side stripes.
DISTRIBUTION Much of Canada and the USA.
HABITAT In the northern part of its range
hibernates for much of the year. Sometimes
climbs low bushes to get to birds' nests.
FOOD Amphibians, worms, the odd rodent and
baby birds.
BREEDING In communal hibernation sites these
snakes can have an extraordinary mating
system. In spring mating balls are formed with
up to 30 males trying to mate with one female.
Some males exude the same pheromones from
their skins as females, so they distract the other
males in the ball and have a better chance to
mate. Live bearers, producing up to 15 young.

SMOOTH SNAKE
Coronella austriaca

The rarest snake in England, but common
throughout much of the rest of its range, even
taking up residence in gardens.
DESCRIPTION Usually up to 2 ft/60 cm. It has a
small head, small eyes with round pupils and a
cylindrical body. Usually gray but grading up to
reddish, with a slightly darker head, a strong
stripe from the neck through the eye and a
series of irregular small dark spots on the back.
DISTRIBUTION Much of western Europe except
southern Spain, much of Britain and northern
Scandinavia. To the east it continues into
Russia, Asia Minor and northern Iran.
HABITAT Diurnal, though secretive. Prefers dry
habitats like sandy heathland, bushy slopes and
embankments.
FOOD Relies heavily on a diet of lizards, though
will also take small snakes, mammals and
insects.
BREEDING Live-bearing; 2–15 young in a litter.

EGG EATING SNAKE
Dasypeltis scabra

Renowned for its ability to swallow eggs far greater in diameter than that of the snake itself.
DESCRIPTION Up to 3½ ft/105 cm. A thin snake with a surprisingly small head, considering its diet. Normally brownish, but occasionally gray or black, with a black "chain" of diamond-shaped markings along the back. The scales are heavily keeled on the back and serrated on the sides.
DISTRIBUTION North East Africa, southern Arabia, west to Gambia and down to South Africa.
HABITAT Partially arboreal, even using birds' nests it has just robbed as resting places. It is found in a variety of habitats apart from rain forest or desert. When molested this snake mimics the defensive actions of venomous snakes, either rasping its scales or puffing up, hissing and feigning aggressive strikes.
FOOD Solely feeds on eggs, using not teeth but projections from its vertebrae to break open the swallowed egg before regurgitating the drained shell.

BREEDING Up to 18 eggs in a clutch.

WESTERN HOGNOSE SNAKE
Heterodon nasicus

Famed for its defensive displays, which range from bluff through to full-scale death-feigning.
DESCRIPTION Up to 3 ft/90 cm. A heavy-bodied snake with thick neck and distinctively upturned nose. Colors range from cream to brown, with heavy light to dark brown blotching along the body and characteristic black markings on the underside of the tail.
DISTRIBUTION Southern Canada ranging south in a wide band through much of central USA and down into northern Mexico.
HABITAT Prefers open land, prairies, sparse woodland, farmland, floodplains and into semi-arid and canyon areas. Uses its broad snout to burrow and its enlarged back teeth in holding prey.
FOOD Mildly venomous. Toads are this snake's staple food, but it will also take frogs, salamanders, lizards, snakes and reptile eggs.
BREEDING A clutch of 4–23 eggs is laid.

EASTERN HOGNOSE SNAKE
Heterodon platyrhinos

Like the Western hognose this snake may provide clues as to how venom evolved; though it has enlarged rear teeth and a very mild venom, it has no real mechanism for introducing this venom to prey.

DESCRIPTION 1½–4 ft/51–115 cm. A stocky snake with a distinctly less upturned nose than its western counterpart. The ground colors are variable, ranging from yellow through gray, brown to even red; some individuals are plain black or gray, but spotted specimens are the norm.

DISTRIBUTION Much of eastern and central USA.

HABITAT Active during the day it spends most of its time foraging for prey, often burrowing into root systems to locate toads. Like the Western Hognose, it will feign death as well as inflate the neck and strike.

FOOD Mildly venomous. Toads form the majority of its diet, but frogs may also be eaten.

BREEDING Lays a clutch of 5–61 eggs.

SCARLET SNAKE
Cemophora coccinea

An impressive mimic of the eastern coral snake when seen at a distance, easier to identify if you can get a closer look.

DESCRIPTION 1¼–2½ ft/36–82 cm. Distinguished by having markings that form a saddle pattern rather than a banded pattern, a very pointed, red snout and a plain whitish belly.

DISTRIBUTION USA, in many of the eastern seaboard states north to New Jersey, down to Florida, through to Texas and north to Montana.

HABITAT Found in loose-soiled open woodland. A burrower rarely found at the surface, more normally disturbed under logs or by agricultural practices.

FOOD A constrictor, feeding on small mice, lizards and snakes, and proving very partial to snakes' eggs.

BREEDING Females lay 3–8 leathery and elongated eggs in a clutch.

SMOOTH GREEN SNAKE
Opheodrys vernalis

The green dorsal color of this snake changes to a dull blue or gray after death.

DESCRIPTION 1–2 ft/30–65 cm. A slender plain green snake with a white or yellowish belly grading up to bright yellow under the tail.

DISTRIBUTION Occurs in much of northeastern USA and some parts of southern Canada as well as isolated populations in southern Texas and Idaho, New Mexico and Wyoming.

HABITAT Mostly terrestrial in grassy areas in forests, prairies and along river edges.

FOOD Eats mainly insects and spiders.

BREEDING Several females may share a nest site where each will lay a clutch of 3–18 eggs.

MANGROVE SNAKE
Boiga dendrophila

A spectacular animal that is a popular "draw" in snake charming shows.
DESCRIPTION Grows up to 8 ft/2.5 m. A glossy black snake with 40–50 sulphur yellow bars. The eye has vertical pupils like a cat's.
DISTRIBUTION Thailand and the Malay peninsular, Philippines and Indonesia.
HABITAT Mangroves and tropical rain forests, where it is often found in the trees.
FOOD Small mammals, birds, eggs and reptiles. Large fangs at the back of its mouth can deliver quite a potent venom.
BREEDING Lays a clutch of eggs in damp soil or rotting wood.

BOOMSLANG
Dispholidus typus

When angered, this snake inflates its throat to produce an alarming threat display to potential predators.
DESCRIPTION Averages 4–5 ft/120–150 cm. It possesses a very short head with large eyes and a slender body. The color varies markedly, even within the same geographic location.
DISTRIBUTION Africa, south of the Sahara.
HABITAT Totally arboreal, the name boomslang comes directly from the Afrikaans meaning "tree snake".
FOOD One of the most notorious and venomous snakes in Africa, back-fanged but because of its short head the enlarged fangs are in fact relatively near the front of the mouth. It preys on chameleons, other lizards, amphibians and birds, often found raiding weaver bird colonies.
BREEDING Up to 24 eggs deposited that require about 6 months' incubation before the 30cm (1ft) hatchlings emerge.

GOLDEN FLYING SNAKE
Chrysopelea ornata

One of the 5 species of snakes that can spread their ribs and glide from tree to tree.
DESCRIPTION Grows to about 4 ft/1.3 m. An athletic, slender snake with large eyes. Its coloration is green with each scale bordered and bisected by black.
DISTRIBUTION India and Sri Lanka, Indonesia, southern China and the Malay peninsular.
HABITAT A denizen of tropical rain forests, it is diurnal, arboreal and sun loving. Moves with alacrity in the tree tops by climbing, jumping and gliding. Back-fanged, but they have a weak venom.
FOOD Lizards and frogs. May take an hour or more to subdue a lizard.
BREEDING Comes down to the ground to lay eggs in leaf mold on the forest floor.

VINE SNAKE
Oxybelis aeneus

May attempt to mimic the branches they habitually lie along by appearing to sway in the breeze.
DESCRIPTION 3–5 ft/90–150 cm. An incredibly slender and long-headed snake, with comparatively small eyes and a long tail (up to half its body length). Generally grayish-brown above, gray below, with white or yellow under the head, an eye stripe and distinctive cream lips.
DISTRIBUTION Extreme South Arizona into Central and South America.
HABITAT Active during daylight hours, mainly arboreal, often being found along thin branches, in a range of arid to moist habitats. Bluffs when disturbed with a wide-gaping mouth.
FOOD Back-fanged and mildly toxic, feeds mainly on lizards.

BREEDING A clutch of 3–5 eggs is laid in spring and summer.

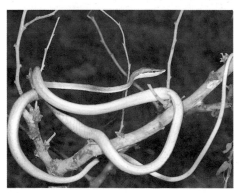

SEA KRAIT
Laticauda colubrina

Can be found in large numbers when they come
ashore to breed.
DESCRIPTION Up to 5 ft/1.5 m in length. Its
coloration is bluish gray with cross bands of
black. They head is marked with black and
yellow.
DISTRIBUTION Tropical seas and shores from
India through Indonesia, Malaysia, New
Guinea, Australia and the Pacific Islands.
HABITAT Mainly aquatic, but also found on the
land in rocky and coral crevices. Venom is toxic
but this snake has a placid disposition and does
not bite even when freshly caught.
FOOD Forages at night, grabbing sleeping fish
from rock crevices.
BREEDING It comes ashore to lay eggs.

FIERCE SNAKE
Parademansia microlepidota

The land snake with the most potent venom. A
large one has the potential to dispatch 250,000
mice.
DESCRIPTION 6½ ft/2 m is the average length.
Usually a brown snake, with some of its scales
edged with black or brown. Some populations
have a distinct black head.
DISTRIBUTION Australia in western Queensland,
northwestern South Australia and western New
South Wales.

HABITAT Found in stony deserts or dry flood
plains with deep cracking clays and soils.
Abroad in daylight, it often lives in the burrows
of its main prey, the plague rat.
FOOD Rodents.
BREEDING Lays a clutch of 9–12 eggs which
hatch in about 70 days.

FIERCE SNAKE

COLLETT'S SNAKE
Pseudechis colletti

Potentially fatal to people, this rare and
beautiful Australian snake keeps itself to itself
and there is no proof that it has ever bitten
anyone.

DESCRIPTION Total length is about 8 ft/2.5 m.
The snake has a brown or black body color,
with blotches of orange or red that merge
together at the sides.

DISTRIBUTION Only found in central
Queensland, Australia.

HABITAT Generally near rivers in black soil
flood plains or riverine forests. It is so
uncommon and elusive that little is known
about its behaviour.

FOOD Small mammals, lizards, frogs and birds.

BREEDING Lays 7–14 eggs in October to
December.

EASTERN TIGER SNAKE
Notechis scutatus

The venomous snake that is common in the most densely populated areas of Australia.

DESCRIPTION A bulky snake that can grow to nearly 6½ ft/2 m in length. Very variable in color, can be gray, green, brown or even black with a series of lighter cross bands.

DISTRIBUTION East and south eastern Australia.

HABITAT Usually found in damp habitats. Unaggressive, but holds its ground, so it can be trodden upon accidentally; before an anti-venom was developed it was responsible for human fatalities.

FOOD Specializes in frogs, but will take birds and rodents.

BREEDING Males indulge in "ritualized combats" during the spring. After mating the female retains her eggs within her body, eventually giving birth to 30 or so young.

BLACK TIGER SNAKE
Notechis ater

Some island populations of this snake fast for 10 months of the year.

DESCRIPTION There are a variety of subspecies that range in size from 3–8 ft/1–2.4 m. All are heavily built, with broad heads. The color is dark brown or black, with some of the western populations having lighter bands.

DISTRIBUTION Western Australia, southern Australia, Tasmania and small islets and islands off Australia's southern coast.

HABITAT Found in marshlands, sand dunes or dry rocky deserts. Some of the island forms spend most of their time in sea bird burrows.

FOOD Frogs, birds and rodents. Two of the island subspecies survive by feeding upon mutton birds, a type of shearwater. The snakes must gorge on chicks when the shearwaters are breeding, then fast for 10 months, when the mutton birds are out at sea. Juvenile tiger snakes live on lizards until they are big enough to eat a bird.

BREEDING Females give birth to 6–20 young (sometimes more) in mid to late summer.

TAIPAN
Oxyuranus scutellatus

The largest and most notorious venomous
snake in Australia, with the longest fangs
(½ in/12 mm) to boot.

DESCRIPTION Can grow to 11½ ft/3.5 m but
more usually 5 ft/1.5 m. It has large glittering
eyes set in a creamy head, with a body of
light to dark brown becoming lighter at the
sides.

DISTRIBUTION North and northeastern
Australia.

HABITAT Found in sugar cane fields, farms,
rubbish dumps and woodlands. If given the
chance a taipan will always retreat from people
but if provoked will strike repeatedly.

FOOD A rodent specialist, which is why it is
attracted to farms and dumps where there are
large populations of mice.

BREEDING Lays 10–12 eggs.

DEATH ADDER
Acanthophis antarcticus

Perfectly camouflaged, it wiggles the tip of its tail to lure unwary prey.

DESCRIPTION Fat body, usually under 3 ft/1 m in length. Very variable in color; red, gray or brown with cross bands that can be either darker or lighter than the overall ground color.

DISTRIBUTION The whole of Australia except the southeastern corner.

HABITAT Forests or scrubland with sand or leaf litter in which it can hide. A "sit and wait" predator that flicks its tail to lure inquisitive prey within striking distance.

FOOD Reptiles, small mammals and birds.

BREEDING A live-bearer that produces a litter of 15–20 babies.

KING BROWN SNAKE
Pseudechis australis

Under threat in the northern part of its range from the lethal effects of swallowing cane toads.

DESCRIPTION Up to 6½ ft/2 m long. Each scale can be edged or tipped with black, resulting in a reticulated pattern on an all-ground color of copper or brown.

DISTRIBUTION The whole of Australia, except the south and the east.

HABITAT Found in nearly every type of habitat from tropical forests to arid desert. In cool weather this snake is active during the day, becoming nocturnal in hotter seasons or climates.
FOOD Small mammals, birds and frogs. For the snake, cane toads seem a perfect food except that they are toxic and eating one causes death. Native animals have yet to come to terms with an introduced species.

BREEDING Ritualised combat has been observed between males during the breeding season (October and November). After mating the female lays about 10 eggs.

CORAL SNAKE
Micruroides euryxanthus

Despite possessing a potent venom, this snake generally defends itself with sound; waving its tail above its body and extruding its cloaca produces "popping" sounds.
DESCRIPTION No more than about 20 in/50 cm long. The body is totally encircled by clearly defined glossy, black, yellow and red bands. The edges of the scales are tipped black, while the head is black from the snout to just behind the termination of the mouth.
DISTRIBUTION Northern Mexico into New Mexico and Arizona.

HABITAT Found in areas of sandy soil in rocky locations emerging from burrows or under rocks at night and during overcast conditions.
FOOD Venomous, feeding almost entirely on snakes, especially the blind snake, *Leptotyphlops*.
BREEDING 2–3 eggs laid under a rock or in a burrow.

EASTERN CORAL SNAKE
Micrurus fulvius

Highly venomous but inoffensive, will rarely bite except under extreme circumstances.
DESCRIPTION Up to around 3 ft/90 cm. The yellow bands are narrow and border the black and red bands; there is some flecking of black in the red bands. The black on the head reaches only to just beyond the eyes.
DISTRIBUTION Southeastern USA, and from southern Arkansas west into Texas and south into Mexico.

HABITAT A secretive snake, it often remains hidden in leaf debris or burrows, only emerging into its woodland or riverine habitat on humid or overcast days.
FOOD Venomous, feeding on small prey items, snakes, lizards and nestling rodents.
BREEDING Deposits up to 18 eggs in rotten logs or stumps.

BUSHMASTER
Lachesis muta

It is reputed that female bushmasters actively and aggressively guard their nest sites.

DESCRIPTION Normally 7–8 ft/210–240 cm, but sometimes up to 12 ft/350 cm. The ground color can vary between yellowish, reddish and gray-brown, marked down the back by a series of dark brown blotches that stretch down the sides to form dark lateral triangles.

DISTRIBUTION Much of Central America and into Peru, the Guianas and into Brazil.

HABITAT Resides almost exclusively in primary and secondary forests and adjacent cleared areas. Mostly nocturnal, foraging for prey on the ground, in burrows and around exposed root systems.

FOOD Venomous, most normally feeding on small mammals, but occasionally eating birds and amphibians.

BREEDING Lays about 12 eggs in a clutch.

SOUTH AFRICAN SPITTING COBRA
Hemachatus hemachatus

"Rinkhals" is the Afrikaans name, referring to the distinctive white throat band.

DESCRIPTION A large stout cobra that is dingy black or brown. Averages about 3 ft/1 m in length. The only cobra with keeled body scales.

DISTRIBUTION Zimbabwe and South Africa.

HABITAT Found in a variety of habitats where it hides in scrubby vegetation or rock piles. When defending itself it can spit venom for up to 3 metres (10ft).

FOOD Rodents and toads.

BREEDING A live-bearer, which is unusual for a cobra. It gives birth to a litter of 63 young in the autumn.

MOZAMBIQUE SPITTING COBRA
Naja mossambica

The "red spitter" can spray two jets of venom from tiny holes in the tips of its fangs for up to 2.5 metres (8ft).

DESCRIPTION Usually attaining a length of 5 ft/1.5 m but on occasions reaching up to 9 ft/2.8 m. Its coloration ranges from brown-red, pinkish to orange-red; there are often black cross bands or blotches on the throat.

DISTRIBUTION Southern Tanzania, Mozambique, Botswana and northern South Africa.

HABITAT Ground dwelling, sheltering in termite mounds and rock crevices, it basks during the day and forages at night. Quick to rouse and to spit, it will also bite readily, though the venom rarely causes fatalities.

FOOD Venomous, eating toads, snakes, birds, rodents and even eggs.

BREEDING Between 10 and 22 eggs are laid in summer.

FOREST COBRA
Naja melanoleuca

Though a slender snake it is the largest of the African cobras.

DESCRIPTION Up to 8 ft/2.5 m. It appears black unless in good light, when a distinct pale flecking of the scales is obvious. The front of the snout, parts of the face and the underside are a bright orange-yellow; on the belly this is broken by a black band under the hood.

DISTRIBUTION Central Africa, south to Angola and eastern parts of South Africa.

HABITAT Occurs in heavily forested areas or along forest edges.

FOOD Venomous, small mammals.

BREEDING Lays 15–26 large eggs in leaf litter or hollow logs.

KING COBRA
Ophiophagus hannah

The largest venomous snake in the world with gigantic fangs long enough to penetrate the hide of an elephant and enough venom to kill it.

DESCRIPTION Usually 13 ft/4 m in length, but the maximum ever recorded was over 18 ft/ 5.5 m. Medium to dark brown with dull light and dark cross bands along the body.

DISTRIBUTION India, Indo-China to southern India, and the Indo-Australia Archipelago.

HABITAT A denizen of tropical rain forests. For most of the year shy and wary of human beings, but females can become aggressive when guarding the nest. A king cobra rearing up over 3 ft/1 m off the ground in a threat display can make for an exciting encounter.

FOOD Rodents and reptiles, with a predilection for other snakes.

BREEDING The only snake in the world that builds an elaborate nest. The female lays up to 40 eggs, staying on guard until they hatch.

MONOCLED COBRA
Naja naja

India's good snake, a valuable exterminator of rats and mice.
DESCRIPTION A large specimen would attain 6½ ft/2 m in length. Typically yellowish or dark brown, with a black and white spectacle marking on the neck that is only obvious when the snake is aroused and the hood is spread. There is also a pair of black and white spots on the undersurface of the hood.
DISTRIBUTION Southern Asia and Indo-Australian archipelago.
HABITAT Found in a wide range of habitats; forest, farmland, even towns. Fast and agile it slips away at the footfall of people. Revered in the mythologies of India and Egypt.
FOOD Mice, frogs and reptiles.
BREEDING Lays 10–20 eggs.

BLACK MAMBA
Dendroaspis polylepis

With a top speed of 23kmh (14mph), this snake is reputedly the fastest in the world.
DESCRIPTION A very large snake up to 14 ft/ 430 cm long. Seldom, if ever, black; they are generally olive-gray to mottled brown with a satiny sheen. The interior of the mouth is, in fact, the only part of the snake that is black. The head is slightly flattened at the sides, making it appear heavier than in many other species from side on.
DISTRIBUTION Africa, from southern Ethiopia down to the Cape.
HABITAT Diurnal and mostly terrestrial, but will climb to bask and search for prey. Said to be as fast and agile in branches or undergrowth as on open ground, it must forage extensively to support such vigorous activity. It has an incredibly high rate of digestion, helped by the potency of the venom, and can digest a large rat in under an hour.
FOOD Highly venomous, it kills prey with a neurotoxin that in human beings could cause death by asphyxia. Eats rodents (including squirrels), hyraxes, birds, bats and lizards.
BREEDING Lays a clutch of 10–15 white oval eggs, usually in a burrow.

GREEN MAMBA
Dendroaspis angusticeps

A regular resident of the roofs of a variety of buildings, including outhouses.

DESCRIPTION 6 ft/190 cm. The body color is normally green with a hint of gold between the scales; sometimes the whole body may be of a yellow hue. Like the black mamba it has a long, angular head, but the lining of the mouth is white.

DISTRIBUTION Throughout much of East Africa below the Sahara and down to the Cape in the south.

HABITAT An arboreal snake, shy and rarely seen in its preferred habitat of well-forested areas, but more regularly encountered in the sparser vegetation of dry bush areas and coastal scrub.

FOOD Venomous, feeding almost exclusively on birds and mammals.

BREEDING Up to 10 eggs are laid in a tree hollow or in leaf litter.

BANDED KRAIT
Bungarus fasciatus

Extremely toxic, but so disinclined to bite that Malayan villagers believe it is not a venomous species.

DESCRIPTION Up to 6½ ft/2 m in size. It has black and yellow cross bands along its entire length.

DISTRIBUTION India and the Malay archipelago.

HABITAT Forests, plantations and farms, where it hides during the day in burrows under stones or logs, becoming active at night.

FOOD Like all kraits other snakes are the preferred food.

BREEDING The female lays eggs.

VIPERS (Family *Viperidae*)

About 190 species. Arguably the most advanced snakes, a group which includes the rattlesnakes, the European Adder (*Vipera berus*) (page 215), and the Gaboon Viper (*Bitis gabonica*) (page 211), which holds the record for length of fangs. This is possible because, as in all vipers, the fangs can be folded into the roof of the mouth and are swivelled forward during a bite. Sometimes the pit vipers (rattlesnakes and their kin) are placed in a separate family. All of them have a heat-sensitive pit on each side of their face, for detecting the body warmth of mammals or birds, enabling them to strike accurately in pitch darkness.

FER-DE-LANCE
Bothrops atrox

More correctly known as Barba Amarilla, this snake has a virulent venom and is reputedly responsible for the most snake-bite related deaths in South America of any species. The true fer-de-lance occurs only on the island of Martinique.

DESCRIPTION Averages 6½ ft/2 m. One of its many common names refers to it as a "lancehead"; its triangular head with light stripes behind the eyes produces an arrow point marking toward the snout. Mostly some shade of brown with darker brown triangles radiating off the back.

DISTRIBUTION From Mexico down through Brazil.
HABITAT Found in plantations and forest, especially along streams.
FOOD Venomous, feeding on small mammals and birds, frogs and lizards.
BREEDING A large female can produce an astonishing litter of up to 70 young measuring 30cm (12in).

GABOON VIPER
Bitis gabonica

This large viper has massive fangs of up to 2 in/5 cm, a record for any snake. The fangs are folded against the roof of the mouth, but are raised and extended when the snake goes to strike.

DESCRIPTION The largest and fattest of the puff adders at up to 6 ft/1.8 m and 26 lbs/12 kgs. Very heavy-bodied, with a wide head and silvery eye marked out by a triangle of brown or black extending from the jaw. The colors of the amazingly cryptic body pattern range from brown, beige, yellow, black and purple.

DISTRIBUTION Much of eastern, central and western Africa.

HABITAT Found on the floors of rain forests and woodland, its remarkable patterning camouflages it wonderfully in leaf litter.

FOOD Venomous, eating a variety of terrestrial species, including some birds and, exceptionally, the small royal antelope.

BREEDING Live-bearing, up to 60 young in a litter.

PUFF ADDER
Bitis arietans

Male puff adders may be seen performing strange ritualized combats or dances together during the breeding season.

DESCRIPTION A very heavy-bodied snake, perhaps only 3–4 ft/1 m, but extremely variable in color, ranging from yellow-brown through reddish brown or gray with heavy black markings.

DISTRIBUTION Africa south of the Sahara, southwest Arabia and Yemen.

HABITAT Most active at night, but in fact generally an ambush feeder, waiting for its prey to pass its silent, camouflaged form. Usually hisses very loudly when threatened.

FOOD Venomous, eating mainly rodents which are quickly digested with the aid of the tissue-destroying nature of the venom.

BREEDING Able to produce massive litters of over 80 live offspring.

CANTIL SNAKE
Agkistrodon bilineatus

The newly hatched young are able to attract their prey by using extraordinary movements of their bright yellow tail lure.

DESCRIPTION A blue-back to chocolate brown snake, with creamy-white markings taking the form of two narrow lateral stripes and regular thin cross bars.

DISTRIBUTION Through much of southern Mexico, Guatemala, El Salvador, Nicaragua, Honduras and Belize.

HABITAT Mainly nocturnal, it generally resides near water.

FOOD Venomous, feeding on amphibians, fish, mammals and reptiles.

BREEDING Live-bearing, giving birth to over a dozen young.

CANTIL SNAKE

HORNED DESERT VIPER
Cerastes cerastes

A snake of shifting sands that moves by "sidewinding".
DESCRIPTION Average length is about 2 ft/60 cm. The color of the snake generally matches the sand surface in the region an individual occurs, varying between pink and yellow, with regular blotching on the back and heavily keeled scales. The most remarkable features are the long horns over the eye that allow sand to accumulate while keeping the eye clear of cover as the snake waits in ambush.
DISTRIBUTION Through much of North Africa and well into the Sahara, to the Middle East.
HABITAT This snake avoids the extremes of desert temperatures by burying itself along the length of its body with rhythmic muscle contractions, then waits for suitable prey to pass by.

FOOD Venomous, feeding on desert rodents and lizards.
BREEDING Lays a clutch of eggs in disused burrows or under stones.

MALAYAN PIT VIPER
Enhydris plumbea

Locally known as the "axe snake" because a bite can have the same result as a blow from an axe – the loss of a limb.

DESCRIPTION Averages 28–32 in/70–80 cm. Patterned with angular markings of dark brown, edged with black on a reddish-brown background.

DISTRIBUTION Indo-China, Malaya, Sumatra and Borneo.

HABITAT Forests and oil palm plantations where it is particularly common. A perfectly camouflaged "sit and wait" hunter that lies motionless even when approached. Very dangerous to "rubber tappers" if they work without shoes.

FOOD Rodents, frogs and reptiles.

BREEDING After laying 13 to 30 eggs, the female coils around them until they hatch about 40 days later. It is unusual for vipers to show maternal care like this.

ADDER
Vipera berus

The adder has the greatest terrestrial range of any snake in the world.

DESCRIPTION Usually up to 2 ft/65 cm but occasionally up to 3 ft/90 cm. The ground color is most commonly gray but may be reddish brown, yellow, olive or greenish. It is most easily distinguished by the dark zigzag pattern running down the back and by the dark V or X shaped mark on the head.

DISTRIBUTION Most of Europe (absent from Ireland, southern Spain, Italy and the southern Balkans) up to the Arctic Circle, east to the northern Pacific coast of China.

HABITAT Very varied, moors, woodland, marshy meadows; it is even a capable swimmer.

FOOD Venomous. It eats small mammals and lizards.

BREEDING Live-bearing; litters vary between 4–12 according to the size of the female.

OTTOMAN VIPER
Vipera xanthina

This snake can be distinguished from the other vipers in its European range, as it is the only one that lacks a nose horn.

DESCRIPTION Up to 4 ft/120 cm. Thick-bodied with no nose horn or characteristic head pattern. The color is variable, gray, sandy or darker, with pronounced eye stripe and a mark on the mouth just under the eye. The dark brown stripe on the back is irregular, often broken up into blotches; the underside is grayish, but yellow or orange under the tail.

DISTRIBUTION Turkey, through Asia Minor into Lebanon and the ex-Soviet Union.

HABITAT Found in open woodland, rocky hillsides, pastures and often cultivated areas. A sluggish viper, diurnal normally, but can be active at night during the hotter months.

FOOD Venomous with a bite that could be fatal to human beings. Feeds mainly on mammals and birds but may take lizards also.

BREEDING Live-bearing, averaging around 15 in a litter.

COPPERHEAD
Agkistrodon contortrix

Though painful, the venomous bite of the copperhead rarely causes fatalities.
DESCRIPTION 2–4½ ft/61–134 cm. Usually a fairly chunky snake. Often distinguished by a coppery-red head and a classic hourglass pattern across the back that is dark brown on the tan, orange or grayish ground color. A pit viper, it has small facial pits; the body scales are weakly keeled.
DISTRIBUTION Much of southeastern USA, except Florida, and bounded in the west by central Texas and Kansas.

HABITAT Found in a variety of habitats from rocky hillsides to lush swamp vegetation, its coloration makes it inconspicuous, but if disturbed it will vibrate its tail rapidly and will strike swiftly.
FOOD Venomous, eating mainly small mammals, but also lizards, snakes, amphibians and insects such as cicadas.
BREEDING Live-bearing; the young have bright yellow tail-tips that fade as the snake ages.

NOSE-HORNED VIPER
Vipera ammodytes

Highly venomous, potentially the most dangerous viper in Europe.
DESCRIPTION Up to 3 ft/90 cm long, but more regularly under 2 ft/65 cm. Stout-bodied with a triangular head, males are more often gray and females browner. A clearly marked zigzag striping is normally unbroken on the back, while the underside is grayish to pink with some darker spotting and red, yellow or green under the tail.
DISTRIBUTION A southern species, from

northern Italy through the Balkans into Greece and South West Asia.
HABITAT Prefers dry, sunny, rocky slopes with some vegetation. Mostly terrestrial, but can climb; usually encountered during the day and when disturbed hisses loudly.
FOOD Venomous. Feeds mostly on small mammals, birds and lizards.
BREEDING The females bear live young, which are born in late summer.

NOSE-HORNED VIPER

COTTONMOUTH
Agkistrodon piscivorus

This dangerous snake will often give warning when disturbed by vibrating its tail and gaping its mouth to reveal the "cotton"-white interior.
DESCRIPTION 2½–6 ft/76–189 cm. A large snake that is dark, olive, brown or black above. The cross banding on the back is darker still, but sometimes hard to see, while the belly is usually a little lighter than the back. Care is needed to distinguish it from water snakes, but the presence of facial pits and (on a dead specimen) the single anal scale are diagnostic.
DISTRIBUTION The southeastern states of the USA, from Virginia in an arc bounded by the eastern seaboard taking in Florida through to Oklahoma and Texas.
HABITAT Semi-aquatic, found in swamps, lakes and ditches. A fairly lethargic snake whose behavior in retreating slowly or standing its ground when disturbed marks it out from the fast-feeling water snakes, as does its habit of vibrating its tail.

FOOD Venomous, mainly feeding on fish, but will take birds, mammals and amphibians, as well as baby alligators and turtles.
BREEDING Live-bearing; gives birth between August and September to up to 15 young. Breeds mainly every other year.

SIDEWINDER
Crotalus cerastes

Famous for its classic mode of locomotion that never allows too much of the body to touch the burning desert surface at any one time..
DESCRIPTION 1½–2½ ft/43–82 cm. A shortish, stubby rattlesnake with prominently rough scales and triangular horns over the eyes.
DISTRIBUTION Northwestern Mexico, southern parts of Utah, Arizona, Nevada and eastern California.
HABITAT Spends the day hidden in mammal burrows or beneath low bushes; it emerges into its arid desert habitat mainly at night. Often encountered basking at the side of roads during the day; otherwise elusive, but the parallel J shaped markings it makes in the sand are a distinctive sign of its presence in an area.
FOOD Venomous; eats pocket mice, kangaroo rats and lizards.
BREEDING Live-bearing, producing 5–18 young in late summer or early autumn.

TIMBER RATTLESNAKE
Crotalus horridus

The only rattlesnake in most of northeastern USA, but relatively common only in undisturbed montane areas as it has been persecuted in much of the rest of its range.
DESCRIPTION 3–6 ft/88–189 cm. Large variation between southern and northern populations. The head may be unmarked or with a dark stripe behind the eye, the back may be dark with blotches on the side and forming cross bands nearer the tail or it may have a brownish stripe running down it with chevron-like cross banding. Both show a black tail.
DISTRIBUTION Much of eastern USA, from Maine south to northern Florida, west into
HABITAT Prefers remote areas, wooded hillsides, rock outcrops, swamps and river floodplains. Active between April and October, mostly during daylight but also at night during the summer. In October the snakes may congregate in great numbers at favored hibernation sites.
FOOD Venomous; often waiting perfectly still in order to ambush prey such as squirrels, chipmunks, mice and birds.
BREEDING Females give birth every other year, producing 5–17 live-born young between late August to October.

TIMBER RATTLESNAKE

WESTERN RATTLESNAKE
Crotalus viridis

One of the most aggressive of the rattlesnakes, its bite can be lethal even if treated.
DESCRIPTION 1½–5½ ft/40–162 cm. Very variable over much of its range, but often with two diagonal stripes on the head, one above and one below the eye. It is mainly some shade of brown with darker, regular blotches on the back and sides that thin nearer the tail and almost join to become bands. The lighter colored tail is ringed with black at the base.
DISTRIBUTION Much of western USA, into northern Mexico and some southern parts of western Canada.
HABITAT Mainly crepuscular, preferring rock canyons and scrubby slopes, but often found in agricultural and suburban locations.
FOOD Venomous, preying chiefly on small rodents.
BREEDING 4–21 live-born young are produced from August to October, after mating either in autumn or in March or May.

WESTERN DIAMONDBACK RATTLESNAKE
Crotalus atrox

Noted for its defensive position when it raises its head well above the coils in a classic S shaped pose and intermittently compounds its threat by rattling its tail.

DESCRIPTION 3–7 ft/86–213 cm. A large snake with variable coloration, usually brownish with pale-bordered diamond-shape patches on the back. The tail is ringed with black and white bands.

DISTRIBUTION Southwestern USA and into northern Mexico.

HABITAT Prefers dry or semi-arid areas like canyons and scrubby plains, but also montane locations and river bluffs. A secretive snake, but one that is often in close proximity to areas of human population and is annually persecuted in the "rattlesnake round-ups".

FOOD Venomous, feeding on rodents and birds.

BREEDING 4–25 young are live-born in late summer.

BLACK-TAILED RATTLESNAKE
Crotalus molossus

Possibly the rattlesnake that is least likely to rattle when disturbed.

DESCRIPTION 2–4 ft/71–125 cm. The body color varies from grayish-brown to a very impressive yellow-brown. The markings on the back are inevitably darker than the ground color and with paler centers; they form diamonds nearer the head and thick bands towards the tail. The pattern is usually edged in white or gray and the tail is plain black.

DISTRIBUTION From Arizona, east into central Texas and south into northern Mexico.

HABITAT Found on cliffs and rock outcrops, often near streams, but also encountered in pine and deciduous woodland.

FOOD Venomous, feeds on a variety of small rodents.

BREEDING 3–6 live-born young are born in late summer.

MASSASAUGA
Sistrurus catenatus

Differs from all other rattlers by having 9 large scales at the front of the head.

DESCRIPTION At only up to 40 in/100 cm, a short, but well-proportioned, snake. A light gray to gray-brown snake with a row of large brown, gray or black blotches down the back and smaller and fainter spots on the side. Has a broad dark eye stripe and a long mark from the head to the neck, sometimes shaped like a lyre.

DISTRIBUTION Southern Canada (Ontario), southwest to Arizona and northeastern Mexico.

HABITAT Prefers moist situations like swamps and around rivers, but in the west it adapts to drier conditions.

FOOD Venomous, eating lizards, snakes, small mammals and frogs.

BREEDING Live-bearing; a litter of 2–19 is born between July and September.

BUGS & BEETLES INDEX

SPIDERS INDEX

Acari 81
Achaearanea tepidariorum 101
African golden orb-weaver 111
Agelena labyrinthica 107
Agelenopsis aperta 107
Amaurobius 99
Amblypygi 82
American green lynx spider 129
American nursery-web spider 127
Aname 87
ant spider 131
Anphaena accentuata 131
Aphonopelma chalcodes 89
Arachnida 81
Araneae 84
Araneomorphae 84, 86, 90–149
Araneus 112–14
Arctosa 126
Argiope 102, 115–6
Argyrodes 102, 111
Ariadna bicolor 95
Arizona black hole spider 100
arrow-shaped thorn spider 123
arthropods 80–1
Atrax robustus 87
Atypus affinis 86
Australian trap-door spider 87

baboon spider, common 89
basilica spider 123
bird-eating spiders 88–9
black lace-weaver 99
black-striped orchard spider 110
black widow spider 103
black zipper 133
blunt-spined kite spider 120
Brachypelma smithi 88
bridal-veil lynx spider 129
bronze Aussie jumper 148
brown spiders 91
buzzing spider 131

Camaricus formosus 144
Carolina wolf-spider 125
Castianeira 130
cellar spiders 92, 103
Cheiracanthium erraticum 130
chelicerae 82–3, 86
Christmas spider 122
Chrysso 104
cobweb spider 108
cobweb-weavers 101–4
comb-footed spiders 101–4
conical orb-weaver 118
crablike spiny orb-weaver 121
crab spiders 83, 138–44
cribellates 83, 97–8
cross spider 112
Cyclosa 118
Cyrtophora hirta 117
daddy long-legs spiders 92
dainty platform spider 106
Deinopsis 96
desert bush spiders 93
desert grass spider 107

dewdrop spiders 102, 111
diamond spider 141
diet 82–3
Diguetia canities 93
Dolomedes 128
Drapetisca socialis 106
Drassodes 133
dwarf spiders 105–6
Dysdera crocata 94

earth-chaser 124
egg-sac 83
Enoploqmatha ovata 102
Eris aurantius 146
European fishing spider 128

feather-footed spiders 97
fishing spiders 128
flower spider, common 142
four-spot orb-weaver 113
funnel-weavers 107–8
funnel-web tarantulas 87

garden spider 112
Gasteracantha 119–122
giant huntsman 135
glasshouse spiders 101
golden orb-weaver 111
golden-silk spider 111
gold leaf crab spider 143
goliath tarantula 88
grass-head sac spider 130
grass spider 139
grass spiders 107–8
great golden argiope 115
green meadow spider 134

habitats 80
hairy tent-spider 117
Hallowe'en crab spider 144
Harpactira gigas 89
harvestmen 81
Hasselt's spiny spider 120
house crab spider 138
huntsman spiders 134–5

invisible spider 106

jaws *see* chelicerae
Johnson's jumper 148
jumping spiders 145–9

knobbly crab spider 144
Kukulcania arizonica 100

lace-weavers 98–9
Latrodectus 103
leaf jumper 147
leaf lace-weaver 98
leopard spider 95
Leucauge 109–10, 123
lichen huntsman 135

long-horned orb-weaver 119
Loxosceles 91
lumpy thorn spider 122
Lycosa carolinensis 125
lynx spiders 129
Lyssomanes viridis 147

mabel orchard spider 110
marbled orb-weaver 114
Mecynogea lemniscata 123
Metellina segmentata 110
Mexican red-knee tarantula 88
Miagrammopes 97
Micrathena 122–3
Microlinyphia pusilla 106
Micrommata virescens 134
missing sector orb-weaver 118
Misumena vatia 142
mites 81
molts 81
money spiders 105–6
mouse spider 132
Mygalomorphae 84, 86–9

Nephila 102, 109, 111
Neriene peltata 105, 106
net-casting spiders 96
Nigma puella 98
nursery-web spiders 127–8

Opiliones 81
opisthosoma *see* abdomen
orb-weavers 112–23
orchard spiders 110, 123
Oxyopes schenkeli 129

painted ground spider 130
palps *see* pedipalps
Pandercetes gracilis 135
pantropical jumper 146
Pardosa 125
Peucetia viridans 129
Phidippus johnsoni 148
Philodromus dispar 138
Pholcus phalangioides 92
Phrynarachne rugosa 142
Physocyclus 92
pigmy mesh-spinners 98
Pirata piraticus 126
pirate spider, common 126
Pisaura 127
plain crab spider 140
platform-web spider 105
Plexippus paykulli 146
prey 82–3
prosoma *see* cephalothorax
purse-web spiders 86

recluse spiders 91
red and silver dewdrop spider 102
red and white cobweb spider 102

sac spiders 130–1
Salticus scenicus 145
sand-runner 126
Scorpiones 81

scorpions 81
Scotophaeus blackwalli 132
Scytodes thoracica 90
Segestria senoculata 95
Selenops radiatus 136
shamrock spider 113
silver argiope 116
six-eyed spiders 94
six-spotted fishing spider 128
snake-back spider 95
Solifugae 81
Sphodros rufipes 86
spinners 83
spitting spiders 90
spotted wolf-spider 125
Steatoda grosa 103
Stephanopis altifrons 144
stick spider 97
stone spiders 132–3
Sydney funnel-web spider 87
Synema 143

tailless whipscorpions 82
tarantulas 88–9
Tegenaria 92, 108
Telamonia dimidiata 149
Tetragnatha extensa 109
Texas orb-weaver 113
Thanatus formicinus 141
Theraphosa blondi 88
Theridion melanurum 101
Thomisidae 83
Thomisus onustus 141
thorn spiders 122–3
Tibellus oblongus 139
triangular spine-leg spider 104
Trochosa terricola 124
tube-web spiders 95
two-striped gaudy jumper 149
two-tailed spiders 137

Uropygi 81–2

variable jumper 146
velvet mite 81
venusta orchard spider 110
violin spider 91

wall crabs spiders 136
warty bird-dropping spider 142
webs 83
wedding-present spider 127
Western desert tarantula 89
whipscorpions 81–2
window lace-weaver 99
windscorpions 81
wolf spiders 124–6
woodlouse spider 94

Xysticus gulosus 140

yard spider 108

zebra spider 145
Zelotes apricorum 133
Zuniga magna 131
Zygiella x-notata 118

SNAKES INDEX